The End of Final Causes in Biology

Lucas John Mix

The End of Final Causes in Biology

palgrave
macmillan

Lucas John Mix
Museum of Comparative Zoology (MCZ) 410A
Harvard University
Cambridge, MA, USA

ISBN 978-3-031-14016-7   ISBN 978-3-031-14017-4   (eBook)
https://doi.org/10.1007/978-3-031-14017-4

This Palgrave Macmillan imprint is published by the registered company Springer Nature Switzerland AG.
The registered company address is: Gewerbestrasse 11, 6330 Cham, Switzerland

PREFACE

This is not a book I intended to write. It arose organically from three research projects that ended up being related. One had to do with the history of definitions of life and the perennial problems of biology. Another related to the problem of detecting long-term trends in evolutionary biology. A third asked about the role of prediction in biology, astronomy, and astrobiology. How do we speak scientifically about the future?

Historically, philosophically, and scientifically we cannot understand where we are going without some idea of where we are and where we have been. Bioteleology deals with the way we talk about the ends of living things including power (to do), information (to be), and purpose (for the sake of which things occur). Many theories of bioteleogy, casually "bioteleologies," have been proposed over the centuries and most have been ridiculed as unscientific. And yet, the historical, philosophical, and scientific reasons behind past rejections are rarely taught, making it difficult for modern biologists to say, in a succinct and compelling way, why these theories remain out of bounds.

I found myself writing a book for new biologists that would help address this. It takes ideas like intelligent design and orthogenesis seriously *in context* to make it clear why they were rejected and why they remain unproductive avenues for biologists.

I am deeply grateful to numerous individuals and institutions that made this book possible. David Haig and the Harvard Department of Organismic and Evolutionary Biology provided support and access to libraries, including the amazing Ernst Mayr Library. David Wilkinson and St. John's

College at Durham University provided time for research. The final stages of the book were written during my time as the Baruch S. Blumberg NASA/Library of Congress Chair in Astrobiology, Exploration, and Scientific Innovation. Resources at the Library of Congress including the archive of Carl Sagan's papers proved helpful.

The central research was conducted as part of the project, "A Long-Term Arrow of Protein Evolutionary Time," with PI Joanna Masel at the University of Arizona. That project was supported by the John Templeton Foundation. (The opinions expressed in this publication are those of the author and do not necessarily reflect the views of the John Templeton Foundation.) The project included an international workshop on Long-Term Trends in Evolution. Special thanks are due to Joanna Masel, Dan McShea, Roberta Millstein, Frank Rosenzweig, Josef Ubeda, and the other participants for their insights.

This work reflects years of discussions and arguments about bioteleology. I am mindful of helpful conversations with Rika Anderson, Andrew Berry, Fred Bookstein, Ford Doolittle, Joe Felsenstein, David Haig, Carlos Mariscal, and John Wakeley. Each of them would no doubt disagree with me on particulars, but I hope I have provided a solid foundation for further discussion.

Cambridge, MA   Lucas John Mix

CONTENTS

CHAPTER 1

An End to Ends?

Abstract Life is dynamic and partially predictable, requiring life-specific language and metaphors. The fundamental units of biology have changed through time, the entities to which we ascribe agency, power, form, and function. Such "biological agents" have coevolved with ideas of nature and teleology. The history of biological agents provides insight into the philosophical foundations, perennial questions, and future possibilities of biology. It reveals how biology naturalized teleology and finalized nature over the past two centuries. The vegetable souls of Aristotle represent a diverse collection of life concepts that took on mental properties in the Middle Ages. The rejection of such theories led to a proliferation of biological agents in the early modern period, from which genes and natural selection and eventually emerged. Modern biology is defined by a shift from agential metaphors to relational and statistical metaphors. The new agents—genes, organisms, and populations—are best understood pragmatically and contextually. Their teleology is bound by methodological naturalism, local adaptation, and blind chance. This reflects a fulfillment of bioteleology, rather than a rejection. Understanding these limits can aid biologists in understanding the history of the field as well as developing new research programs.

Keywords Adaptation • Chance • Gene • Naturalism • Teleology

L. J. Mix, *The End of Final Causes in Biology*, https://doi.org/10.1007/978-3-031-14017-4_1

Ideas about organisms evolve, much as organisms themselves do. They occur in populations. They persist through time. They vary in multiple ways. And they experience selection based on their specific time and place. This book traces one population of ideas from the beginnings of biology through to the present day.

Biologists attribute power, form, and function to individual actors or "agents." These actors can seem obvious. Most people—biologists included—find the idea of an organism easy to use when thinking about macroscopic plants and animals. And yet, the more we learn about the variety of life, the more complicated the story becomes. A single body can have multiple genomes (e.g., chimeras and symbioses) and a single genome can stretch across multiple bodies (e.g., clones and colonies). Does "organism" refer to the body, a self-contained regulatory unit, or to the genome, a unit of information and fitness? Similarly, "gene" can refer to a short chain of polynucleotides, a batch of information, or a driver of biological action (e.g., "regulating" protein action or "directing" protein synthesis). Which one is the real gene? And how literally should we take talk about power, form, and function. Genes do not "act" in the same way that humans do, but gene action is key to biology.

No textbook can set forth the ideal definition for these terms. They are not perfect, eternal categories, cleaving nature at the joints. They are useful descriptors that change with time and context. I have chosen to treat "gene" as a species of ideas that displays variation across interpretations of power, form, and function. Collectively, these three terms capture the teleological or end-oriented features of living things: the ends themselves (function), the actor who has them (form), and the ability to achieve them (power).

The evolutionary approach to a history of ideas views "gene" as a collection of related ideas. The population of ideas varies, and some are more adaptive than others. Some do more work or communicate more clearly about biological processes. And thus, the population shifts. The "gene" has evolved from older ideas, each with its own variation and context. My goal, then, is to trace the history of these ideas in a way that reveals which aspects of them are adaptive. The conceptual environment (biology as a natural science) made biological agent concepts what they are today. It constrains the ways biologists attribute power, form, and function to them. They, in turn, shaped their environment, reframing "nature" and natural science to accommodate biology.

Tracing the coevolution of concepts and their conceptual environment can help biologists better understand both historical controversies and contemporary limits in biology. It reveals three key lessons. First, older biological agent concepts (e.g., vegetable souls and vital particles) did not simply die away; they evolved into contemporary concepts, contributing both language and theories for understanding power, form, and function. Second, some concepts died out because they were—and still are—maladaptive. Modern "genes" and "organisms" each describe a range of ideas, but that range is limited. Bioteleology now relies on evolutionary theory, an explanatory recursion of form and function, and a well-characterized network of biochemical reactions. History explains function for ideas as well as organisms.

BIOLOGICAL ACTORS

Whales differ dramatically from their terrestrial ancestors. Horses have grown larger over millennia. And humans have larger brains than early hominids. An evolutionary approach to ideas begins with a recognition that the current definition of a term will be different from earlier definitions. And yet, they are connected by descent.

Power, form, and function as categories have a deep history, being descended from the efficient, formal, and final causes of Aristotle. Chapters 2–4 trace that history, looking at what they meant then to better understand what they mean now. I introduce them here to set up the argument of the book, but the concepts themselves are no more eternal than species of whales, horses, and humans. I understand them evolutionarily *because* they change through time. If I have done my job, you will think of them differently by the time you reach the final chapter.

Function and Teleology

Biologists speak of function very carefully. Dictionaries define function as a purpose, what something is supposed to do. It may invoke design, intention, or expectation, but it invariably appeals to a proper or correct end for something. The car is *for* driving. The heart is *for* pumping blood. Biologists often distinguish between a biological "function"—imparted by evolutionary or metabolic context—and a more anthropomorphic "purpose"—imparted by human intent. The former answers "what for?" without suggesting a designer, a mind, a prediction, or moral progress. This seems to be necessary for a good explanation in natural science.

Biology requires "what for?" explanations. But is a scientifically respectable version possible? Over the past three centuries, biologists have explored numerous constraints that limit how they talk about function and other ends—such as health, fitness, and adaptation. Biologists settled on three constraints, discussed in Chap. 7: methodological naturalism, local adaptation, and blind chance. The counter-intuitive shape of those constraints warrants an extensive literature in theory, philosophy, and history of biology—and an introductory book.

In this book, I use the term teleology very broadly to refer to "what for?" explanations or ends language. That includes biological concepts of function and fitness. It also includes appeals to personal intent, supernatural design, and retroactive causation. Whenever a proper end is invoked, it counts as teleology. The language of ends has remained the same through the centuries as has the idea that ends explain something critical to biology. Interpretations have changed, what the language means and how the explanation works.

I speak more narrowly of function as a proper end of a biological structure or behavior. It answers a particular "what for?" question. Pumping blood helps explain the shape and activity of a heart. Producing a new oak tree helps explain the chemical composition of acorns. Gene regulation helps to explain methylation in nucleic acids.

This book describes many different theories of biological teleology or "bioteleologies." Chapters 1–6 are descriptive. They follow historical agents and bioteleologies without endorsing them. Chapters 7 and 8 are normative. They argue that biologists can still explore a variety of bioteleologies within the three constraints I describe. Biology requires teleology and multiple options exist.

Form and Agents

Form may be the most challenging of the three traits to understand. It is alien to modern ears and less clearly teleological than power and function. It is, nonetheless, an essential starting point. Any invocation of ends requires some notion of who or what has the end. The easiest example, and the one most often used, refers to minds and mental states. The automotive engineer has an idea of what she wants when she designs a truck. The "standard story" or standard model of agency begins with an abstract mind holding abstract beliefs that somehow cause changes in the physical world (Hornsby 2004). Numerous philosophers have challenged this standard story, but it remains the default model after the Enlightenment.

Bioteleology depends on some theory about biological individuals capable of having ends. Natural selection, for example, requires units of selection, entities fitted (more or less) to their environment. Biologists speak of these units *as though* they had minds capable of holding beliefs and causing changes. They speak as though organisms "want" to survive and reproduce, as though genes "act" to perpetuate themselves.[1]

Biologists from Aristotle onward recognized that many, perhaps most, organisms lack the kind of mind present in the standard story. How, then, should we understand the language of ends? No easy answer presents itself. Aristotle spoke of (natural) vegetable souls (Chap. 2). Medieval thinkers exported biological ends to the (supernatural) mind of God (Chap. 3). Early modern biologists attempted to fit ends back into nature by reimagining both biological units and nature (Chap. 4). By the twentieth century, genes, organisms, and environments emerged as the relevant loci for ends, albeit in different ways for different biologists. Biological function and fitness relate to the ends of survival and reproduction for genes and organisms. Thus, genes and organisms are said to have interests and act on them even when they do not have minds. The environment, meanwhile, acts on genes and organisms through natural selection.

In this book, I use the term biological agent to refer to the "for whom?" of bioteleology. That includes Aristotle's vegetable souls, Mendel's genes, and a menagerie of strange entities in between (e.g., vital spirits, *moule intérieur*, and *gemmules*). It is the intellectual genus of which gene and organism are extant species. By the end of the book, I will show that they are useful indices for understanding a continuous process, pragmatic categories justified by evolutionary theory.

---

[1] See Mix 2018, pp. 8–10, for a defense of the term "agency" when speaking of natural, biological activity. Due to the Enlightenment divide discussed below, modern English forces a distinction between unnatural mental agency and natural physical passivity. Biology requires a third space of "proto-agency" causally distinct from either and many terms have been proposed. The circumlocutions necessary to describe biology with entirely passive language are prohibitively difficult, perhaps impossible—I have tried. As this is a book on teleology and the term "agent" forces reflection on precisely those issues at stake (power, form, and function), I have used it for the non-standard agents of biology. I do not intend to sneak in any ontology. To the contrary, my hope is to stimulate critical thought on how biology handles both teleology and ontology.

I use form to describe the defining characteristics of biological agents, what makes them what they are, especially when those characteristics are hidden. Biologists have long sought to understand how an acorn "knows" to grow into an oak tree. The form of a tree is hidden in the acorn. Saying that information resides in the acorn's genome does not make concepts of agency and form go away. Instead, it assigns agency to the genome and frames form as information.

The physical shape of the tree is encoded in the genes, but it does not physically reside there. There is no tiny tree shape present in the polynucleotides. Neither do modern biologists appeal to an immaterial Platonic form residing eternally in the mind of God. Encoding, then, is an inherently teleological concept that requires a natural biological agent to translate shape into code with the "intent" that another biological agent will translate the code back into a physical shape. This is, of course, only the most trivial example—many traits are not morphological. Form describes the aspect of biological traits that can be hidden away. That includes the defining traits of an agent, but also by extension, other types of information stored by or for that agent.[2]

Power

One question remains. How do the "for whom?" biological agents accomplish the "what for?" biological ends? The standard story of agency requires not only mind and belief, but causal efficacy. Agents have power to achieve ends. They are part of a causal narrative. Standard agency has rarely been considered an option and so biologists work—and have always worked—to imagine non-standard agents for their explanations.

[2] It may seem that current models of "information" avoid both agency and teleology and, thus, are meaningfully distinct from Aristotelian forms and formal causes. This is too lengthy a discussion to pursue here, but I would add two notes. First, etymologically, the word "information," like the word "organism," came into English as an inherently agential term. In the fifteenth century it was a verb ("to give advice") and thus a noun ("a piece of advice"). Only in the twentieth century did it take on a sense of "a mathematically defined quantity divorced from any concept of news or meaning." The *OED*, curiously, cites R.A. Fisher in 1925 as the first use of this latter definition (c) but attributes it to Claude Shannon in 1948. Notably, Shannon entropy is alphabet/language dependent. It requires knowledge of the repertoire of possible states and, therefore, presumes both an agent (to encode) and a teleology (to communicate). "information, n.". OED Online. June 2022. Oxford University Press. https://www-oed-com.ezp-prod1.hul.harvard.edu/view/Entry/95568?redirectedFrom=information& (accessed June 20, 2022).

I use power to describe the causal efficacy attributed to biological agents. Discussions of genes "regulating" metabolism and organisms "encoding" information entail some idea that the physical world is different than it would have been if the agent had not acted. I will argue in Chap. 8 that genes and organisms lack the power attributed to earlier biological agents. Evolutionary theory places power in a dynamic system including environment and population. It cannot be localized to an individual gene or organism. In this way, current bioteleology differs from previous theories.

Broadly speaking, biological agents are the units of biology. When we tell a story about life, they are the individual characters. When we diagram a biological system, they are the nodes. The units shape the stories we tell, even the kinds of stories we can imagine.

These are the entities to which we attribute biological function and activity. In a more remote way—by analogy, extension, or emergence— they are the entities to which we attribute fitness, intention, preference, purpose, and meaning. Thus, they include modern genes and organisms as well as historical entities such as souls, vital spirits, *gemmules*, *ids*, and a host of other proposed difference makers in biology.

Each of these ideas was sincerely intended and fruitfully used to describe life within the social and conceptual bounds of natural science at one time. Most have been supplanted by genes, organisms, and populations because those agents were better suited to biology. They allow biologists to tell better stories, to make better predictions, and understand life better.

THE ARGUMENT

This chapter sets the context for the overall argument of the book: that "gene" concepts are best understood as descendants of older ideas. Life is dynamic and partially predictable, requiring life-specific language and metaphors. Modern biology is defined by a shift from essentialist metaphors (e.g., souls, spirits, and vital particles) to relational and statistical metaphors (e.g., populations and gene types). And yet, the new metaphors retain both the language and the conceptual structure of agency and ends. Ultimately, this reflects a fulfillment of bioteleology, rather than a rejection.

Natural science in the modern vein began with an explicit exclusion of standard agency. The Scientific Revolution partitioned the universe into natural and agential realms, matter and mind. Natural objects took up space, but remained utterly passive, without power, form, or function.

Agents existed in an ideal realm of thought and "acted" upon matter, giving it motion, meaning, and purpose.

Whether viewed as ontologically independent (Descartes), epistemically independent (Kant), or miraculously joined (Leibniz), teleology was excluded from "nature" and placed firmly outside the purview of science. Human agency was ruled spiritual or psychological and non-human agency dismissed as non-sensical. Thus, all non-human agents could be safely ignored: souls, spirits, sprites, fairies, ghosts, gods, angels, and so on.

This fully mechanical view of nature was never truly embraced. Philosophers and scientists continued to look for explanations of the human-like, agency-like features of animals, plants, and other living things. Biologists held out from the mechanical consensus into the early twentieth century as many questioned the viability of a mechanical understanding of life. Philosophers balked at the unbridgeable gap introduced between humanity and the rest of nature.

From the seventeenth to the twentieth century, biologists attempted to reintroduce biological agents that could fill the gap. Such concepts placed power, form, and function back within the realm of nature, requiring new, more natural views of agency and new, more agential (and more teleological) views of nature. These concepts proliferated, mutated, and spread. Some were selected, while others went extinct. Biologists developed a series of rules, excluding "vitalist" and "teleological" accounts of life. New models arose to describe the unique power, form, and function of organisms.

This book covers three distinct constraints on teleology within biology: methodological naturalism, local adaptation, and blind chance. Each constraint does important work. The remaining teleology of form and function was applied to genes as the new biological agent, while the teleology of power, the ability to make a causal difference, was shifted to population-level processes, including natural selection.

I begin with the deep history of biological agency. Chapter 2 addresses the perennial questions of biology, including what distinguishes living things from other objects and what it means for organisms to persist through changing material. Questions of power, form, and function have always been central. They reflect Aristotle's efficient, formal, and final causes, which, when the same, defined the "vegetable soul" or dynamic principle of life for most of Western history.

The next two chapters follow the evolution of vegetable souls from Aristotle to Descartes with an eye toward the de-naturalization of teleology. As final causes came to be viewed as intentions for matter, rather than descriptions of processes in matter, they took on unnatural overtones that would make them problematic for natural scientists. Chapter 3 covers the gradual separation of psychology and physiology in Antiquity and the Middle Ages. Chapter 4 focuses on the final divide of physics and metaphysics in the Enlightenment which rendered vegetable souls non-sensical and created a need for new biological agents.

Chapter 5 looks more closely at the menagerie of biological agents in the early modern period, the purposes they served, and the ways they contributed to current understandings. During the seventeenth, eighteenth, and nineteenth centuries, Aristotelian souls speciated into diverse biological actors, including preformed embryos, vital particles (e.g., *gemmules, plastidules,* and *autocatalytic enzymes*), vital forms (e.g., *moule intérieur* and *orgasm*), and vital forces (e.g., *mesmerism, galvanism,* and *élan vital*). Viewing these changes as adaptations reveals the conceptual history of biology and the ways that empiricism, mechanism, evolution, and natural selection shaped modern science. Reciprocal accusations of "vitalism" reflect competition over who would define nature and natural science. A variety of teleologies were embraced and rejected, leading to a new biological actor, the gene, discussed in Chap. 6. Natural, context-specific, and blind, its agency depended on interactions with other genes and with the environment.

Chapter 7 surveys three philosophical adaptations that enabled modern genes to arise. Each is cast as a constraint on the evolution of teleology in biology, an exclusion of one way of thinking about ends. Methodological naturalism limits evolution to the lawful regularity of natural causes. It excludes intelligent design and other appeals to unnatural ends as unscientific. Local adaptation shows that the course of evolution is usually divergent. Long-term directional trends can occur, but most proposed trends have been disproved, including progressive evolution or the "path to perfection." Blind chance describes the reformulation of biological actors and environmental forces to exclude prospect or future imagination. Orthogenesis and Lamarckism were rejected, but their elements were reformed into stochastic, population-based theories of change.

In Chap. 8, I look more closely at how evolutionary theory harmonized teleology and nature. Genes and natural selection provide an etiological recursion of form and function, each justifying the other as populations travel through time, adapting to their environments.

The title of this book, *The End of Final Causes in Biology*, represents my own ambivalence about teleology. On the one hand, biologists have made significant advances by carefully excluding three kinds of teleology. On the other hand, this work has culminated in a new kind of teleology informed by evolutionary theory. This mature teleology provides insights into the unique directedness of genes, organisms, and populations. The rejected bioteleologies should still be actively resisted by biologists and philosophers alike, not because they are ideologically forbidden or abstractly false, but because they are concretely maladaptive. They decrease the fitness of theories by rendering them less elegant, less comprehensible, and less predictive. Meanwhile, new theories should be proposed that build on historical, philosophical, and experimental insights.

THE CONTEXT OF THE ARGUMENT

This is a littoral book, existing in the tidal zone between the dry land of biology and the depths of the humanities, specifically the history and philosophy of biology. No doubt the denizens of one will find it a bit squishy, the inhabitants of the other a bit shallow. This is always the case working across disciplinary boundaries. My primary goal is to tackle teleology for early career biologists who wish to understand how the field got where it is and where it might be going. They can get their feet wet.

I hope some will dive deeper. Each chapter includes detailed discussion and references, but the overall context invites comment. Teleology has proven one of the most controversial issues in the philosophy of biology, with extensive commentary by biologists, philosophers, and historians, each with their own perspective and focus. For the adventurous, I have included a few notes to place my argument in academic context. More casual readers may wish to skip ahead to Chap. 2.

Biological Context

My primary discussion partners are and have been biologists, among whom David Haig has had the greatest influence. His 2020 book *From Darwin to Derrida: Selfish Genes, Social Selves, and the Meanings of Life* tackles the evolutionary theory aspects of this discussion with greater sophistication than is possible here. I am largely convinced by his approach to biological agents: strategic genes as heritable difference makers. "Genes have a causal role in the production of bodies, and bodies have a causal role in determining which genes survive the filter of natural selection" (230). The causal question cannot be separated from this recursive process.[3] Nor can we make sense of genes either as independent physical tokens or as comprehensive types, the set of all identical tokens. We must approach genes as a collaboration between some, but not all, copies operating in a historical population. I take from Haig the etiological priority of a population undergoing natural selection and a pragmatic approach to biological categories and language.

I believe that Haig underestimates the philosophical weight of language, readily using terms like information and agency without appreciating the connotations smuggled in, even in his own writings. For example, he defines information by means of observers and interpreters, then describes ribosomes as "mindless interpreters" (244–5). It remains unclear to me what observation and interpretation mean absent a mind to internalize (i.e., sense or observe) and process (i.e., interpret) previously external forms. This has been a persistent question in the history of biology, specifically at the traditional boundary between vegetable and animal life (Mix 2018, pp. 239–248). I am doubtful that the concept of interpreter can be saved in a world without minds. Worse, I doubt the comprehensibility of any truth claim in such a world. What is truth if not a belief about the world, accurately representing it, but meaningfully distinct from it, existing in the mind of an interpreter? Nonetheless, as an evolutionary theorist I am satisfied that Haig has described a sufficiently natural picture of biological agents and teleology. A pragmatist, he has sidestepped the ontological questions of objectivity and truth, providing a very useful theory. This may be the best option available within the context of natural science.

My approach is both humbler and broader. It is humbler in that I am not attempting a single definitive interpretation of biological agency and teleology, much less of human agency and teleology, as Haig reaches by the end of his book. Rather, I am attempting to set forth the broad outlines for agency and teleology *within* modern biology. This territory encompasses the strategic gene but leaves room for alternative explanations that also build on evolutionary theory.

---

[3] This causal recursion is not original to Haig. In the context of twentieth-century evolutionary theory, it owes much to Richard Dawkins' (1976) *Selfish Gene* and both draw on George Williams' (1966) *Adaptation and Natural Selection*. Haig (2012) introduces his specific "strategic gene" approach.

More widely, this book reflects my attempts to interpret and internalize the teleological insights of Ernst Mayr (1961, 1988, Mix 2016) and other proponents of the Modern Synthesis. I have sought to respect their contributions without succumbing to a type of adaptationism that claims exclusive knowledge of evolutionary causes.

Orzack and Forber (2017) provide a brief survey of adaptationism, distinguishing empirical, explanatory, and methodological flavors (following Godfrey-Smith 2001). A strict *empirical* adaptationism acknowledges adaptation as the only significant causal factor in evolution. Other factors are negligble such that any change which cannot be understood as adaptation cannot be understood. *Explanatory* adaptationism suggests that understanding adaptation is the primary goal of evolutionary biology (as in Dawkins 1976 and Dennett 1995). *Methodological* adaptationism makes a less ambitious claim that adaptation is a good organizing principle and should be the first tool used in understanding evolutionary change. I am an adaptationist in this last, admittedly weak sense. However defined, I suspect that Gould's repeated attacks on adaptationism—as well as Dawkins' (and Haig's) defense—drove them to stronger philosophical positions than were necessary for their biological arguments.

Debates around adaptationism frequently reflect a desire to move from pragmatic claims about research priorities to ontological claims about the nature of causation. In biology, discussion of "constraints" tends to contrast the passivity of environmental conditions with the agency of natural selection. I am sympathetic with a focus on selection, but not with the causal characterization. Precisely because I favor natural explanations for biological phenomena, I am unwilling to privilege the agency of selection. More importantly, I remain skeptical of researchers' ability to identify the relevant targets of selection in all cases. Adaptationist explanations are often beyond our reach, and this need not derail biology.

Eldredge and Gould (1972) and their successors in paleobiology have convinced me that there are interesting and tractable research questions for which the adaptationist paradigm is inappropriate. Sepkoski (2012) surveys the investigation of evolutionary trends in paleobiology, while Felsenstein (2004) discusses stochastic elements of modeling evolution in phylogenetics. No doubt other examples could be found as well.

Philosophical Context

I am indebted to Denis Walsh for his philosophical work in this area, most substantially his 2015 volume, *Organisms, Agency, and Evolution*. Walsh provides invaluable insights into the complex relations between genetic accounting and phenotypic accounting. He deftly tackles the interweaving of historical, philosophical, and scientific issues. At the same time, he reifies organisms in a problematic way. I share his concern that biologists privilege genes to the exclusion of other explanatory frameworks. And I share his concern that evolutionary biologists uncritically dismiss processes occurring within an organism. And yet, I remain unconvinced that his proposals will address these issues in a meaningful way. "Organism" turns out to be as problematic to define as "gene." Perhaps more so. From cancer to colonies to the genet-ramet distinction, biologists have reasons to doubt that organism is a natural kind or even a sufficiently unambiguous category for analysis (Godfrey-Smith 2009, pp. 69–86). Rather, there exist multiple, non-exclusive, and overlapping biological units worthy of consideration (Mix 2018, pp. 225–238). Each can display power, form, and function in its own way. Each competes and cooperates with similar entities within a population against a background environment. More specifically, Walsh's repeated description of genes and traits as "sub-organismic" seems to be begging the question. Godfrey-Smith (2009) and Haig (2020) provide reasons for thinking this too narrow an understanding of the role that genes and traits play at the population level. The benefits of organism-based approaches do not outweigh the costs of weakening the philosophical bulwarks of the Modern Synthesis.

A closer read of Walsh reveals an attribution of agency neither to genes, nor to organisms, but to the "organism situated in a system of affordances" (2015, p. 209). Walsh highlights the pragmatic quality of biological units and how they are reciprocally defined with biological environments. He credits this reciprocal notion of biological agency—including power and form—to C.H. Waddington, Richard Lewontin, and J.J. Gibson, particularly the latter's theory of affordances (p. 172). His ends, however, while nominally spread across a range of biological units, regularly return to organismal ends and organic function (sub-organismic purpose derived from organismic ends). I return to the question of reciprocal theories of agency and privileged levels of analysis in the final chapter.

More broadly, I am engaging with the ontology of Ruth Millikan and Karen Neander, both of whom use natural selection to understand the identity (formal cause) and function (final cause) of biological agents. In the process of defining "proper function" historically, Millikan (1984, 1989) and Neander (1991) develop an etiological understanding of genes and organisms based in a history of adaptation. To identify the function of a thing (a biological agent) using the history of selection on it is to identify the thing itself as a product of evolutionary history. And not just evolution generically, but specifically the selective events in its lineage. I have attempted to respect this insight and fill it out in a biologically concrete way.

Peter Godfrey-Smith makes an important contribution to this ontology in *Darwinian Populations and Natural Selection* (2009). He looks in greater detail at evolutionary history as a set of population-level processes, asking what persists and what varies. In this way, he arrives at a less intuitive, but more robust theory of biological agents and agency. While I do not agree with him on all particulars, his theory falls within the bounds described in this book.

Throughout, I have attempted to incorporate the work of two philosophers of biology who pay close attention to the probabilistic processes underlying evolution. Elliott Sober's contributions are too extensive to list, but I recommend *Evidence and Evolution* (2008) and *The Nature of Selection* (2014), which look closely at the logical framework for evolutionary explanations. Sober elegantly describes the constant stochastic background from which adaptation must be distinguished. Roberta Millstein (2000, 2002) asks about the nature of random processes in evolution, particularly those labeled "drift." She makes important distinctions, including one between "drift" construed as a random process and "drift" as a probabilistically modeled outcome. Their work has led me to my conclusion that power must be more clearly articulated at the system level and not localized to genes and organisms.

Historical Context

This volume continues a project begun in my 2018 book, *Life Concepts from Aristotle to Darwin: On Vegetable Souls*. My purpose is to track the transmission of biological theories from one generation to the next. Questions around power, form, and function have remained remarkably similar over three millennia of broadly "Western" biology.[4] This is not to deny important generational differences. Nineteenth-century biologists simply were not asking the same questions, nor considering the same conceptual solutions, as their seventeenth-century ancestors, much less the natural historians of Antiquity. And yet, each generation inherits the language, confidences, and anxieties of their predecessors. The entelechy of Hans Driesch (1867–1941) was not the same as the entelechy of Aristotle (384–322 BCE) and yet Ernst Mayr (1904–2005) read, interpreted, and replied to both when discussing teleology. It bears asking how all three concepts, all three views of biological agents, constrain the imagination of current biologists. *Life Concepts* focused on "vegetable souls," the dominant biological agent concept from Aristotle until the Enlightenment, still discussed by biologists through the nineteenth century. This book has a different conceptual focus—teleology—and a narrower scope—eighteenth- to twentieth-century biology. It looks at the conceptual descendants of vegetable souls, asking which philosophical features they inherited and where mutation occurred.

While many books have been written on the history of biology in the context of modern science, they generally focus retrospectively on links with modern theories. Far fewer look substantially at the way Medieval and Renaissance perspectives were perpetuated. I am happy to recommend Margaret Osler's *Reconfiguring the World* (2010) and Michel Foucault's *Order of Things* (1994) for detail on intellectual history at the beginnings of modern biology. For a closer look at the strong influence of nineteenth-century German biologists, consider Timothy Lenoir's *Strategy of Life* (1982), Elaine Miller's *Vegetative Soul* (2002), and John Zammito's *Gestation of German Biology* (2017). Evelyn Fox Keller (2003), Michael Ruse (2009), and Mary Midgley (2002, 2010) provide important critiques of contemporary biological concepts, including how they retain older, and often problematic, teleology and ontology.

---

[4] In *Life Concepts*, I focused on historical questions and the specific influence of Aristotle's four causes on biological thinking over three millennia. I speak there of "cause, identity, and purpose." In this book, I focus on issues in theoretical biology over the past three centuries and, therefore, use "power, form, and function." There is a strong correlation between the two triads. Power, form, and function are narrower and more explicitly modern categories.

Naturalized Teleology

Modern biology "naturalized" teleology. It brought power, form, and function back from the agential (and thus unnatural) realm of the Enlightenment and placed them in the purview of empirical reasoning. Natural selection and genetics played a large role in the creation of (non-standard) biological agents that could do this work while still following the rules of natural science.

Modern biology also "finalized" nature, stretching the rules of natural science beyond the narrow limits of atomism and passive matter imagined by Descartes. By establishing (natural) biological agents, capable of having interests, encoding and following programs, and regulating chemical processes, biology rehabilitated teleology, if in a narrow way. To miss this second point is to misinterpret the foundations of modern biology (a philosophical error) and misunderstand the role of fantastic beasts such as the *moule intérieur* (a historical error). More important for biologists, it makes it all too easy to fall into old and erroneous ways of thinking.

Humility toward the past should provide humility for the future. While the referents and broad outlines of "organism" have remained the same from the time of Aristotle, the details have changed dramatically, particularly in regard to power, form, and function. The basic category endured, but the meaning drifted. Mendel proposed genes in an abstract and immaterial way. It took several decades before they could be associated with DNA, and biologists still struggle with the relationship between the physical gene (a string of nucleotides, a token) and the strategic gene (a unit of inheritance, a type). No doubt gene concepts will continue to evolve.

The history of biological agents reveals why modern biologists speak of genes as they do, why genes outcompeted Renaissance souls and Romantic vital spirits, and why the conceptual bounds of the nineteenth and twentieth centuries remain important for research. Attempts to introduce additional teleology (e.g., design and progressive evolution) can harm the overall fitness of evolutionary theory. Even scientific proposals (e.g., an expanded synthesis) must overcome well-established challenges.

*   *   *

Over the past few decades, historians have become increasingly wary of the "Scientific Revolution" as a useful descriptor in intellectual history. Critics point to the focus on how the idea spotlights Europe and emphasizes discontinuity with previous perspectives. This is not to deny the revolutionary influence of scientists from Copernicus to Darwin, but to recognize the continuity of rational thought and transformative observation throughout Medieval Europe, but also in Golden Age Islam, India, and China. Following the continuity thesis of Pierre Duhem and others, I do not see a radical break in biological thinking between periods.

In the history of ideas, I am a gradualist and not a saltationist. Significant mutations, even Bauplan innovations, can occur, but we should look for minor changes first and constantly ask how each concept was adapted to its particular conceptual environment.

On a similar note, I have concerns that the "great man" view of science can harm both science and science communication. In his book, *On Heroes*, Thomas Carlyle (1866) credited the genius and power of a few remarkable individuals who set the course of history. Within the sciences, this has led many to underestimate the role of community and consensus building, including data sharing, collaboration, and peer review. More broadly, it has led various publics to expect science to move more swiftly and decisively than empiricism, skepticism, and due diligence demand.

Contrary to popular belief, Copernicus did not immediately revolutionize astronomy. Nearly 200 years passed before Newton provided the necessary concepts to understand planetary orbits. Astronomers debated numerous theories and continued collecting data throughout that period and would have done so, even without social pressure. Nor was Newton's model identical to those of Copernicus and Galileo.

Science takes time and we should not be surprised if evolutionary theory takes 200 years to come to terms with Darwin's insight. Indeed, it would be deeply disturbing if our current theory agreed with Darwin on all particulars. Attempting to shoehorn biology into "Darwinian," "Neo-Darwinian," or "Modern Synthesis" molds reflects too much dependence on the genius of a few. It absolves individual scientists (along with other scholars and publics) of responsibility for understanding the conceptual foundations and limitations of their models. Conversely, knowing those foundations prepares them for active participation in shaping biology.

There are many ways to get the teleology right. It is possible to think and talk about biological ends that are consistent with modern science and available data. It is possible to make bioteleology transparent and productive. It is also possible to get the teleology badly wrong. Biologists benefit from a clear statement of when, how, and why that happens.

REFERENCES

Carlyle, Thomas. *On Heroes: Hero-Worship and the Heroic in History.* New York: J. Wiley, 1866.

Dawkins, Richard. *The Selfish Gene.* New York: Oxford University Press, 1976.

Dennett, Daniel C. *Darwin's Dangerous Idea.* New York: Simon & Schuster, 1995.

Eldredge, Niles, and Stephen J. Gould. "Punctuated Equilibria: An Alternative to Phyletic Gradualism." In *Models in Paleobiology,* edited by Thomas J. M. Schopf, 82–115. San Francisco: Freeman, Cooper & Co., 1972.

Felsenstein, Joseph. *Inferring Phylogenies.* Sunderland, MA: Sinauer Associates, 2004.

Foucault, Michel. *The Order of Things: An Archaeology of the Human Sciences.* New York: Vintage Books, 1994.

Godfrey-Smith, Peter. "Three Kinds of Adaptationism." In *Adaptationism and Optimality,* edited by Steven Hecht Orzack and Elliott Sober, 335–357. New York: Cambridge University Press, 2001.

Godfrey-Smith, Peter. *Darwinian Populations and Natural Selection.* Oxford: Oxford University Press, 2009.

Haig, David. *From Darwin to Derrida: Selfish Genes, Social Selves, and the Meanings of Life.* Cambridge, MA: MIT Press, 2020.

Haig, David. "The Strategic Gene." *Biology & Philosophy* 27, no. 4 (2012): 461–479.

Hornsby, Jennifer. "Agency and Actions." In *Agency and Action,* edited by John Hyman and Helen Steward, 1–24. Cambridge, UK: Cambridge University Press, 2004.

Keller, Evelyn Fox. *Making Sense of Life: Explaining Biological Development with Models, Metaphors, and Machines.* Cambridge, MA: Harvard University Press, 2003.

Lenoir, Timothy. *The Strategy of Life: Teleology and Mechanics in Nineteenth Century German Biology.* Chicago: University of Chicago Press, 1982.

Mayr, Ernst. "Cause and Effect in Biology: Kinds of Causes, Predictability, and Teleology Are Viewed by a Practicing Biologist. *Science* 134, no. 3489 (1961): 1501–1506.

Mayr, Ernst. *Toward a New Philosophy of Biology, Observations of an Evolutionist.* Cambridge, MA: Harvard University Press, 1988.

Midgley, Mary. *Evolution as Religion: Strange Hopes and Stranger Fears.* New York: Routledge, 2002.

Midgley, Mary. *The Solitary Self: Darwin and the Selfish Gene.* Durham, UK: Acumen, 2010.

Miller, Elaine P. *The Vegetative Soul: From Philosophy of Nature to Subjectivity in the Feminine.* Albany, NY: State University of New York Press, 2002.

Millikan, Ruth. *Language, Thought, and Other Biological Categories.* Cambridge, MA: Bradford Books/MIT Press, 1984.

Millikan, Ruth. "In Defense of Proper Function." *Philosophy of Science* 56, no. 2 (1989): 288–302.

Millstein, Roberta L. "Chance and Macroevolution." *Philosophy of Science* 67, no. 4 (2000): 603–624.

Millstein, Roberta L. "Are Random Drift and Natural Selection Conceptually Distinct?" *Biology and Philosophy* 17, no. 2 (2002): 33-53.

Mix, Lucas J. "Nested Explanation in Aristotle and Mayr." *Synthese* 193, no. 6 (2016): 1817–1832.

Mix, Lucas J. *Life Concepts from Aristotle to Darwin: On Vegetable Souls*. New York: Palgrave, 2018.

Neander, Karen. "Functions as Selected Effects: The Conceptual Analyst's Defense." *Philosophy of Science* 58, no. 2 (1991): 168–184.

Orzack, Steven Hecht and Patrick Forber. "Adaptationism." In *Stanford Encyclopedia of Philosophy*, Spring 2017 ed. Stanford University, 1997–. https://plato.stanford.edu/archives/spr2017/entries/adaptationism/.

Osler, Margaret J. *Reconfiguring the World: Nature, God, and Human Understanding form the Middle Ages to Early Modern Europe*. Baltimore: Johns Hopkins University Press, 2010.

Ruse, Michael. *Monad to Man: The Concept of Progress in Evolutionary Biology*. Cambridge, MA: Harvard University Press, 2009.

Sepkoski, David. *Rereading the Fossil Record: The Growth of Paleobiology as an Evolutionary Discipline*. Chicago: University of Chicago Press, 2012.

Sober, Elliott. *Evidence and Evolution: The Logic Behind the Science*. Cambridge, UK: Cambridge University Press, 2008.

Sober, Elliott. *The Nature of Selection: Evolutionary Theory in Philosophical Focus*. Chicago: University of Chicago Press, 2014.

Walsh, Denis M. *Organisms, Agency, and Evolution*. Cambridge: Cambridge University Press, 2015.

Williams, George C. *Adaptation and Natural Selection: A Critique of Some Current Evolutionary Thought*. Princeton, NJ: Princeton University Press, 1966.

Zammito, John H. *The Gestation of German Biology: Philosophy and Physiology from Stahl to Schelling*. University of Chicago Press, 2017.

CHAPTER 2

What Makes Life Life-Like? The Dynamic Continuity of Living Things

Abstract Living things share a strange property recognized from the earliest biological observations: continuity through change. This property underlies the history of terms like organ, organic, and organism, which suggest multiple parts sharing a common purpose, preserving a living thing through nutrition and reproduction. It provides a teleology that remains in modern biology. A gene is a polynucleotide with context and purpose. Genes and proteins reveal a pervasive teleology in biochemistry; they are composed of abiological elements yet make up larger organisms. Aristotle identified fundamental issues of composition (material cause), cause (efficient cause), identity (formal cause), and end (final cause). His vegetable souls describe a confluence of the latter three acting dynamically to direct the first, form informing matter. Aristotle set the stage for two millennia of biological theory including intense discussion about ends, in Aristotle's words "that for the sake of which" a thing occurs. The cyclical quality of Aristotle's definitions provides key insights to biology. Efficient, formal, and final causes remain important in biology as power, form, and function but all three have varied considerably over the centuries, leading to many interpretations of teleology and soul that are inconsistent with modern science.

Keywords Aristotle • Gene • Nutrition • Organism • Teleology • Vegetable soul

L. J. Mix, *The End of Final Causes in Biology*,
https://doi.org/10.1007/978-3-031-14017-4_2

Living things share a strange property recognized from the earliest biological observations: continuity through change. More active than rocks
and streams, organisms grow and move. A rock remains a rock while an
egg hatches. And yet, organisms actively resist change. They react, resist,
and regulate, maintaining their status and shaping their environment. This
dynamic persistence may be the most interesting and unique aspect of
biology. We attribute agency to living things because they have this end
and do work to pursue it.

This chapter explores the early roots of teleology in biology. Ancient
thinkers recognized that living things differed from non-living things
because of the ends they had. Organs share a common purpose, coming
together to form an organism. The function of parts reveals the form of
the whole, just as perpetuation of the whole provides ends for the parts.
Aristotle focused on the end of nutrition, informing matter and incorporating it into a biological body. He used four types of explanation to
describe what makes life life-like: material, efficient, formal, and final
causes. The latter three correspond to power, form, and function as
described in Chap. 1. Together they define, for Aristotle, what it means to
be a living thing. His primary biological agent, the vegetable soul, would
become the basis for biological thinking in Europe for two millennia. The
four causes and souls were interpreted in diverse and changing ways over
the centuries, often taking on ideal, supernatural, and mental overtones
inconsistent with modern science. Aristotle's approach, however, has sur-
prising resonance with modern biology. In particular, he introduced a
causal recursion that fits well with natural selection.

Continuity Through Change

Organisms grow and develop; populations adapt; species evolve and
change. Still, we have no problem identifying them as organisms, populations, and species. Few, if any, inanimate things have the same fluid consistency and we treat them as borderline cases, using the language of
life, even while thinking of them as not-life. We say that winds die down,
crystals grow, and stars evolve. We feed fires and kill viruses.

Specialized scientific language for biological change and continuity can
obscure the deep history and enduring challenges of speaking about life.
Terms like biochemistry, physiology, and metabolism all refer to physical
processes associated with life without identifying what exactly differentiates them from abiotic processes. The difference between chemistry and

biochemistry hangs on teleology. Biochemistry has a purpose: the continuity of a living system. Individual reactions have specific functions in the service of particular biological agents.

The term "organic" displays the complex history of this idea.[1] In Greek, Latin, and Old French, it referred to a tool or implement: an object used with a particular end. It entered English in the fifteenth century and took on a meaning related to biological teleology. The parts of a body, organs, make a up a single living thing, an organism, defined by a shared purpose. Organs were conceived as instruments of a unifying soul. "Organic" disease occurred when parts failed in their service to the whole. Only in the early nineteenth century was the term extended to other organized structures. The verb "organize" shows a parallel development, while the noun "organism" appears to be a derivative of organ arising in the nineteenth century. All three are inherently teleological.

"Organic chemistry" first referred to the chemistry of living things, but it evolved to mean chemistry involving carbon-carbon bonds—common in, but not unique to, living systems. Organic chemistry remains organic whether or not it is biochemical. At some point in the future, life may well be understood such that biochemistry can be defined without appeal to teleology, but for the moment it remains ambiguous. All biochemical reactions follow the same rules as abiotic reactions; the only difference is context. Biochemistry refers to reactions in the context of perpetuating a living system.[2]

Critical distinctions can still be made at the molecular level, larger than atoms but smaller than organisms. There is a difference between a polynucleotide, a single physical molecule, and a gene, a biological difference maker. Setting aside for the moment whether that gene function is objective or merely attributed, we should ask why and how it is attributed (or discerned). A sequence in a database is clearly a gene, but not a polynucleotide, while a synthesized polyadenosine is a polynucleotide, but not a gene. A gene is a polynucleotide with a purpose. It is a biological agent, to which can be attributed power (e.g., to regulate), form (e.g., to

---

[1] "organic, adj. and n." *Oxford English Dictionary*, 3rd Ed. Oxford: Oxford University Press, 2004.

[2] I have avoided saying directly that "biochemistry refers to reactions perpetuating living systems." We can easily imagine maladaptive, dysfunctional, and diseased reactions and pathways. Some perpetuate a second living system such as a virus, cancer, or pathogen, but others do not. Oxidative stress and prions do not have a clear biological end but, when they occur in metabolic context, they may be called biochemical.

code for), and function (e.g., to be fit). Nor are genes alone in this. The terms protein, enzyme, and hormone require biological ends as well. Biochemistry entails teleology.

Ancient Greek thinkers had similar words. Plato (428–348 BCE) referred to the appetitive soul (*psyche epithumetikon*), a vital principal that empowers desire (see Plato's *Republic*; Mix 2018, 34–36; Lorenz 2006). For Plato, all living things, including plants, had an appetitive soul. It has no explanatory power and so fails as a scientific theory, but it identified teleology as key to life in the same way that biochemistry, physiology, and metabolism do today.[3]

Aristotle (384–322 BCE) provided a more useful approach. An ardent biologist, one-third of his writings describe organisms or biological processes, as much as he wrote on politics, physics, or abstract philosophy. Aristotle appealed to a nutritive soul (*psyche threptikon*), the process by which organisms assimilate foreign atoms, in other words food, into themselves (see Aristotle's *On the Soul*; Mix 2018, 55–66; Johnson 2005; Lorenz 2006). For Aristotle, it was more than abstract teleology; it was the primary function of a living being, persisting through time by drawing matter and energy into itself. His nutritive soul was not an immaterial entity which enabled nutrition; it was the physical process itself, in action and in fulfillment. Later interpreters would reframe and Platonize the nutritive soul, but for Aristotle, it may have been quite similar to modern metabolism. What exactly did the work and what it meant for that thing to have a function were problematic but, as we shall see, they remain problematic today.

Plato's ideas and Aristotle's ideas became fused in the Middle Ages as a "vegetable soul" (Latin: *anima vegetabilis*, literally a soul capable of growth). Sometimes it was conceived very Platonically—ideal, immaterial, immortal—and sometimes very concretely—natural, embodied, mortal—but usually somewhere in between. The term "vegetable soul" sounds ridiculous to modern ears because "vegetable" and "soul" belong to two different modes of thinking, often considered incompatible. Mix (2018) follows the vegetable aspect as physiological and psychological accounts of life diverged historically. Those interested in the soul aspect, consciousness and reason, can find more in Martin and Barresi's (2006) *Rise and Fall of*

---

[3] Note that the OED entry for metabolism specifies "The chemical processes that occur within a living organism *in order to* maintain life" (italics added). "metabolism, n" OED, 3rd Ed. 2001.

Soul and Self as well as Goetz and Taliaferro's (2011) *A Brief History of the Soul*.

For 2000 years, vegetable souls (a conceptual species) were the most successful biological actors in Western biology. They can be viewed as descendants of Plato's appetitive souls and Aristotle's nutritive souls. They inherited aspects of power, form, and function from each and the population of vegetable souls (concepts) in the Middle Ages showed considerable variation. They were held together by a recognition that biological agents are best understood through a teleology of nutrition. They are capable of nutrition (a power), defined by nutrition (their form), and aimed at nutrition (a function).

This chapter looks at the perennial questions of biology and what makes life life-like. These questions have not changed from the time of Aristotle, even though the language and conceptual frames have shifted dramatically. They set the stage for debates at the origins of modern biology, debates around power, form, and function. Like contemporary biologists, historical biologists were deeply concerned with the chemistry of metabolism and reproduction, the ways that living things maintain continuity between moments (homeostasis) and between generations (inheritance). We have come to think of these in terms of genetic action, but the vegetable soul was the biological agent of choice for millennia.

THE PARMENIDEAN THREAD

Plato and Aristotle were both responding to a debate that took place a century earlier between Parmenides of Elea and Heraclitus of Ephesus. Heraclitus looked at things in the world, notably organisms and rivers, and argued that everything changes. Nothing stays the same from moment to moment. Everything is always in a state of flux. He famously quipped that you cannot step in the same river twice. Every organism becomes a new species each time it ingests food or excretes waste. Every skin cell sloughed off redefines it.[4] Parmenides countered that no knowledge would then be possible. Instead, he argued, we must believe in eternal entities—atoms or organisms or ideas—that transcend changing appearances. We can only speak of knowledge as insight into something regular enough and durable enough to be permanent. For Parmenides the real things, the knowable

---

[4] This observation is often presented as the Ship of Theseus paradox and remains an active question in philosophy. See Gallois (2016).

things, cannot arise or depart. Nothing comes from nothing, and some-thing will always be something. As painted, both positions are untenable, but the argument reveals an important question about how we think about the world. What changes? And what persists?

Plato and Aristotle invoked vegetable souls as intermediate entities, more than the ever-shifting particles they informed, yet less than the spe-cies they instantiated. Plato's continuum of souls was less dualist than most readers realize, but it fits poorly with modern metaphysics and thus is of minimal interest for modern biologists.[5] In sketch, he viewed souls as immaterial agents that organize the material world. Aristotle's views prove more informative for modern biologists.

Aristotle identified four types of explanation. Usually translated as "four causes" they might also be thought of as four ways to satisfy a question "why is X the way it is?"[6] Each one reveals something interesting about why life is life-like. The four causes are material, efficient, formal, and final. Each has been reinterpreted through the centuries, but the broad catego-ries remain useful for framing key issues in biology.

MATERIAL CAUSE: COMPOSITION

Material cause explanations address composition—what smaller units make up X? Aristotle was, at least in part, an atomist. He thought the world was composed of tiny, indivisible particles of earth, water, air, and fire, which gave sensible matter its properties. Each had consistent proper-ties and consistent motions responsible for the traits of visible matter. The downward inclination of earth made objects heavy, while the inherent

---

[5] "Soul is the overall working out of order, and the soul aspects are individual stages in the process." Mix (2018, 37). For more on this, see Plato's *Timaeus* and Barney et al. (2012). The critical break for modern readers will be the switch from Platonic Realism to Nominalism in the High Middle Ages. For Platonists, concrete particulars are mere shadows of eternal substantial forms. The forms are more real and thus true knowledge will always be knowl-edge of things that are immortal and immaterial. Souls were not forms, but an intermediate, which could literally *inform* particulars. For Nominalists, categories (e.g., animal, mammal, human) are just names used to describe collections of mortal material individuals. Modern science considers true knowledge to involve statements about particular objects and events—data.

[6] Mix (2018, pp. 42–53) goes into greater detail. See also discussion in Haig (2020) and Walsh (2015, pp. 192–194).

motion of fire provided heat and motion. These atomic properties could be used to explain all motion in the abiotic world.[7]

Some contemporaries believed that fire was sufficient to explain the dynamism of living things. Aristotle discussed the role of fire in both breath and blood but felt that something else was needed. This something extra explained the difference between living and non-living beings. One possibility involved a fifth element, more refined and active than fire but distinct from the ether that filled the heavens (Freudenthal 1995). This fifth element may contribute to a theory of life, but it failed to explain several crucial observations.

Significantly, Aristotle recognized a continuity of life through changing material. The elements in an individual organism and the individuals in a species change over time. At the elemental level, material cause explanations could not provide a Parmenidean thread. Something else was needed. Aristotle turned to a position called *hylomorphism*. Substances—the basic units of existence—require both matter (*hyle*) and form (*morphe*).

The story becomes somewhat confusing because Plato thought forms were substantial in and of themselves. Medieval scholars interpreted Aristotle's forms Platonically, making them an additional component, but this violates the four-cause schema.[8] The form cannot be a part, separable or inseparable, of the whole; it must be something else, as we shall see below. Aristotle came to identify life with the agent capable of maintaining the continuity of matter through nutrition, development, and reproduction.

Despite a radically updated view of elements, modern biologists face the same problem. Atoms and molecules flow in and out of organisms day by day, even second by second. Polymerases work constantly to copy and spell-check genomes. Stem cells divide and differentiate. Germ line cells mutate and divide. At the atomic level, everything changes; nothing stays the same. At the molecular level, modern biology begins to deal with persistent, biology-specific molecules, primarily genes, but here we have already taken a step away from material identity and atomic composition. Genes have power, form, and function, making them biological agents and not simply passive polynucleotides.

---

[7] Johnson (2005) explores Aristotelian teleology in detail, including the abiotic inclination of elements.

[8] Aristotle supported substantial forms in *Metaphysics*. For simplicity, I discuss the hylomorphic substances described in *Categories* and discussed in Shields (2007, pp. 53–64). This seems to be the approach Aristotle took for vegetable souls.

Efficient Cause: Power

Efficient cause explanations address the actor or means responsible for an event occurring. Of Aristotle's four causes, it is the only one still spoken of as a cause in modern English. Who or what caused it to be the way it is? Parents explain offspring; predators explain hunting and killing; pathogens explain disease. Equally, mitosis, meiosis, and sex explain offspring; heterotrophy explains hunting and disease. Unlike material causes, efficient causes need not be exclusive. A long chain of individuals, arts, and events can lead up to the current one. You, for instance, are likely the result of two parents, four grandparents, eight great grandparents and a host of other relatives who materially and socially contributed to your life. You are also the product of consuming untold calories of heat and grams of organic carbon provided by animals, plants, and eventually the Sun (itself powered by gravity and the Big Bang). Aristotle did not have the scientific details we do, but he did conceive of a plurality and infinite regress of efficient causes.

Who, you might ask, is ultimately responsible? Where in the network of causes could things have gone differently? Aristotle distinguished between proximate and ultimate efficient causes. We do not blame the cue ball for a missed shot at pool. Its motion was determined by the strike of cue stick and, ultimately, the person shooting. Aristotle spoke of the ultimate cause as an uncaused cause, the first link in a chain. All subsequent events, up to the one in question, are more proximal causes.

Following the Enlightenment, modern thinkers tend to speak of humans as uncaused causes and treat everything else mechanically. Aristotle, however, did not intend organisms or (nutritive) souls as ultimate causes. Aristotelian biology, like modern biology, was content to dwell in the realm of proximate causes. A nutritive soul is a biological agent with power to perpetuate itself but that does not exclude its also being a collection of elements and the offspring of another nutritive soul.

Some causes will, of course, be more interesting than others. This does not grant them any special ontological character. It does not mean they subsist (exist on their own), have standard agency (with prospect, intentionality, and unnatural causal power), or are in other way distinct from nature.[9] These would be anachronistic descriptions. Like the word "organic," the word "agency" carries philosophical baggage from the Enlightenment, and it remains unclear whether we can separate out the ultimate cause (and teleological) valence of words like act, active, action, and agent. More on this later. For now, I will reiterate that Aristotle favored parallel and chained efficient causes and focused on proximate causes in biology.

When discussing living things, Aristotle was particularly interested in the efficient cause of organisms. What actor or means is responsible for nutrition, the incorporation (literally in-body-ment) of elements? Unlike particles, when organisms collide, there is no conservation of individuals. The lion eats a gazelle and only the lion remains (albeit with more mass). What actor or means is responsible for reproduction, the duplication of living things? An atom of Beryllium-8 decays rapidly into two Helium-4 atoms, but an amoeba divides to form two new amoebas (albeit with less mass). Something interesting occurs in biological nutrition and reproduction. This is still labeling and not yet explaining, but it labels something important. Living things have biological activities and, therefore, one can ask about the relevant biological agent. Who or what has the power to do these things?

Aristotle answered that there must be a biological "self" capable of assimilation (feeding itself) and replication (copying itself). Figuring out what that "self" is would require two more causes, the formal and final cause, each of which has proven problematic for modern science.

FORMAL CAUSE: FORM

Formal cause explanations appeal to the pattern or shape of a thing. More abstractly, and more problematically, a formal cause has often been construed as the essential character of a thing, what gives a thing substance or makes it what it is. When new elements are added to a body, there must be a plan for where they go. Nutrition, according to Aristotle, requires

---

[9] See Chap. 1. Hornsby (2004) introduces and critiques the standard model of agency involving prospective imagination and metaphysically distinct causal power. Numerous more natural views have been proposed, particularly for use with non-human actors. Because prospective imagination is ruled out in biology—see Chap. 7—something along these lines will be needed. I use "agency" for the life-specific causal power common to all organisms. In David Haig's (2020, 281) language, it is an element in the causal chain that is a relevant difference maker.

information (literally en-form-ment), the fitting of abiological elements into an organic, functional whole. What are these forms and where do they come from?

Formal causes appear most dramatically in the process of development. A simplistic view of nutrition might posit that an organism replaces atoms in a pre-existing form and needs no template other than the existent body. Likewise, we could imagine reproduction by a one-to-one mapping of the old body onto a new body, complementary mapping (as with DNA repli-cation), or some other form of splitting and extension (as with crystal fragmentation and nucleation). None of these accurately describe biologi-cal growth, however. Many organisms pass through radically and formally distinct stages.

Acorns, somehow, store the information necessary for a fully grown oak tree. Nutrition then must access forms that can be hidden during part of a lifecycle. Aristotle viewed biological development as a gradual revelation of form. He asked how the information for an oak might be present in an acorn, how the information for an adult human might be present in a sperm cell.

Unlike matter, the form and the power of en-form-ment seemed to be consistent through time, uniquely so in living things. Together they defined a living thing. Aristotle described the formal cause as "what it is" (*to ti esti*) and "what it is to be" (*to ti en einai*). Latin translators failed to find another way of saying this and coined the term *essentia* to capture Aristotle's idea. Originally, an essence was neither more nor less than the defining character of a thing, though Aristotle considered the possibility of something more Platonic and later biologists would move it in that direction.

In the discussion of vegetable souls, the process of forms informing matter—through nutrition, development, and reproduction—provides an essence, an identity, a formal cause: what it is to be a living thing. The concept is somewhat circular in that the process of en-form-ment defines the form itself. This may have contributed to subsequent confusion. It anticipates, however, the circular quality of natural selection, as species identity cannot be distinguished from the historical process of adaptation and a continual feedback loop between population and environment.

Aristotle, leaning toward Parmenides and Plato, thought that biologi-cal forms were eternal. The properties of organisms were knowable because they could not change through time. Acorns could consistently grow into oaks because the pattern of oak-ness remains the same forever. Note that

Aristotle's nutritive soul was a mortal biological agent defined by an eternal form; it was not itself an eternal form.

Modern biologists think very differently about the knowability of particulars (via nominalism) and the evolution of forms (via natural selection), but the basic idea still has appeal. We still view genes, organisms, and species as forms informing matter. We still think of their information content traveling through time, despite being composed of atoms that come and go constantly.

Final Cause: Function

This brings us back to teleology and final causes, one of the most debated issues in philosophy. Aristotle described the final cause as "that for the sake of which" something occurs (Johnson 2005, pp. 64–80). It might concretely involve the intention of a human person, but it might also refer abstractly to a future state relevant to the object or event in question. We have no difficulty imagining that acorns are "for the sake of" propagating oaks, despite knowing that they do not always fulfill this function. We have no difficulty describing the proper function of hearts, intestines, and gametes. They serve a purpose in the context of nutrition, development, and reproduction. Aristotle was explicit in stating that this latter form of teleology, without personal intent, also constituted a final cause, though Medieval (and many modern) philosophers debated what this could mean (Johnson 2005). As with the term "organized" discussed above, Aristotle began with biological function. "Paradigm instances of designerless function-laden entities are the parts of animals" (Shields 2007, p. 72). Aristotle attempts to weave a middle way between those who would deny any ends in nature and those who would ascribe all ends to intentional design (68–90).

Nutrition, development, and reproduction all include a final cause for Aristotle—the perpetuation of "self." But here, he did something interesting. Aristotle defined life as a confluence of efficient, formal, and final causes. His nutritive soul described the "self" that is brought about by, defined by, and aimed at the incorporation of matter into a living system. He considered development and reproduction to be characteristic of all nutritive souls as well. Nutrition was fundamental, however, being responsible for the moment-to-moment persistence of a living being that made longer term processes possible.

In explaining his process-oriented biological agent concept—the soul—Aristotle used several terms that would prove problematic for later thinkers. First and foremost, he invoked a *telos*, or end, from which we get the world teleology. A final cause explanation could be summarized as an end, goal, purpose, or most problematically an intention for a living thing. This made sense for human actions but was less clear when discussing plants. In such cases, "that for the sake of which" proved obscure, even in Aristotle's time. Final causes have something to do with a dynamic process. Like Heraclitus' river, the soul cannot be observed statically, it must be observed in action. In Aristotle's words, a dead body is only a biological body equivocally. It is only a lump of matter unless enlivened by biological processes. The soul—as form and organizing principle of a living being—must be absent after death. A dead body is not informed in the relevant way.

> Aristotle set forth his concept of souls in *On the Soul*. He insisted that souls are things in-action (*energeia*) and in-fulfillment (*entelechia*). Few words have been as thoroughly debated as these two. They both have something to do with a distinction between activities that may occur and activities that are occurring. Like Plato, there is a sense that the world is in a state of becoming. Aristotle focused on the becoming and claims knowledge about it, not just about the ultimate end of being. He spoke of potentiality (what may happen) and actuality (that which is happening). Matter is the first potentiality of life. The soul activates the matter, making the soul the first actuality of life. This activated thing is not just matter, but an organized body. It is not, however, an organism. Not yet. The soul is also the second potentiality. The second actuality is a soul in action and in fulfillment, the body taking nutrients, the living being. The end, as stopping point, can never be reached. The organism will never be sufficiently fed for all time, only for now. Plants are mortal. They are fulfilled not by satiety but in continuous nutrition. That is their essence. A thing can be both actual and active. For Aristotle, it was precisely this actual activity that characterizes life. (Mix 2018, 56)

In other words, a vegetable soul is a biological agent brought about by nutrition and reproduction (in an ancestor) but also possessed of nutrition and reproduction as powers. It is defined by nutrition as the process that makes it what it is (gives it form), but it also applies this form to new matter. Finally, its primary function is to continue in this dynamic process of nutrition.

OTHER SOULS

Aristotle suggested that some souls could be defined by additional activities. Although nominally described only in the context of nutrition, these higher order activities warranted further classification and higher order souls. Sensation and willed motion defined the sensitive soul, though again the soul came about in the process of sensing and willing; it was not an immaterial or compositional cause of these activities. Because sensing and willing require a material body, the nutritive processes could be described as instrumental for them, and a single sensitive soul could describe a living being with all these activities. Medieval scholars conflated the sensitive soul with Plato's spirited soul (*psyche thumoeides*) to form the "animal soul" (Latin: *anima sensibilis*, literally a soul capable of sensation).

The classical split between animals and plants comes from this ancient distinction between organisms capable of sensation (and growth) and those capable of growth alone. Confusingly, English inherits the term "animal" from the first part of *anima sensibilis* and the term "vegetable" from the second part of *anima vegetabilis*. Plato, Aristotle, and countless Medieval thinkers—both secular and religious—attributed souls to both.

Finally, Aristotle suggested that humans possessed one more activity, reason, which distinguishes them among the animals. The rational soul (*psyche noetikon*) allows humans to contemplate the reality and the existence of forms beyond the limitations of transient matter, to escape Heraclitus and change completely. Real knowledge depended upon this transcendent faculty and hinted that human souls—though we experience them in the context of bodies, nutrition, and sensation—may be something more. The rational soul, also appearing in Plato (as *psyche logistikon*), would continue through the Middle Ages (Latin: *anima rationalis*, literally a soul of reason) and be the dominant model for modern soul concepts.

Aristotle believed that biology and value are tightly coupled. The highest good of any living thing could be found in the exercise of its highest activity, plants in nutrition and reproduction, animals in sensation and motion, and humans in contemplation. The descriptive aspect of the final cause blended into a normative aspect of *eudaimonia*, usually translated as "flourishing." The higher order souls and value claims would be problematic for modern biology. These issues were only exacerbated by two historical trends, the Platonization of Aristotelian biology and the growing divide between accounts of bodies and minds.

*   *   *

Aristotle provided the basic categories by which we still speak of life and life-like behavior. Drawing on older discussions, particularly the three souls of Plato, he spoke of souls as biological agents, the entities that distinguish life from non-life and biological order from abiological order. The key feature of souls was the identity of efficient, formal, and final causes, the confluence of power, form, and function in a single individual.

Final causes anchored Aristotle's biology and made sense of the complex causal relationship between matter and form in living things. Their centrality in biology for the next 2000 years led to a constant and continual re-evaluation of what "that for the sake of which" really means. The teleological preoccupation of twentieth-century theoretical biologists and philosophers of biology reflects only the most recent outbreak of anxiety about final causes. Power, form, and function identify the strange continuity through change exhibited in biological processes, namely, nutrition, development, and reproduction. Aristotle's contribution involved the necessary and cyclical link between all three as forms informing matter.

A biologist at heart, Aristotle derived his causal explanations from observing plants and animals. He reasoned by analogy to abiological phenomena. Thus, we cannot look to him for a blank slate of abiotic "nature" or "physics" to which biology can be added—or from which it might be removed. This makes it difficult to answer the question "were Aristotle's final causes natural?" His view of nature was inherently biological and teleological. It was not, however, intentional or supernatural in ways that would upset modern biologists.

Several intellectual trends obscured Aristotle's insights and mutated the vegetable soul into something strange. Conflation with Platonic souls allowed them to be abstracted from their material substrate. The final cause, in particular, came to be associated with minds instead of bodies. Meanwhile, speculation on spirituality and subjectivity opened a gap between the physiological accounts of vegetable souls and the psychological accounts of rational souls. The naturalization of plants and the mentalization of final causes resulted in the latter being removed from nature and placed in the mind of God, an idea utterly at odds with empiricism. The next chapter looks at how vegetable souls evolved through Late Antiquity, the Middle Ages, and the Renaissance, a process that would eventually de-naturalize final causes and de-finalize nature. This set the stage for a rejection of "teleology" by the founders of modern science.

REFERENCES

Barney, Rachel, Tad Brennan, and Charles Brittain, eds. *Plato and the Divided Self.* New York: Cambridge University Press, 2012.

Freudenthal, Gad. *Aristotle's Theory of Material Substance: Heat and Pneuma, Form and Soul.* Oxford: Oxford University Press, 1995.

Gallois, Andre. "Identity over Time." In *Stanford Encyclopedia of Philosophy*, Winter 2016 ed. Stanford University, 1997–. https://plato.stanford.edu/archives/win2016/entries/identity-time/.

Goetz, Stewart, and Charles Taliaferro. *A Brief History of the Soul.* Malden, MA: Wiley-Blackwell, 2011.

Haig, David. *From Darwin to Derrida: Selfish Genes, Social Selves, and the Meanings of Life.* Cambridge, MA: MIT Press, 2020.

Hornsby, Jennifer. "Agency and Actions." In *Agency and Action*, edited by John Hyman and Helen Steward, 1–24. Cambridge, UK: Cambridge University Press, 2004.

Johnson, Monte Ransome. *Aristotle on Teleology.* New York: Oxford University Press, 2005.

Lorenz, Hendrik. *The Brute Within: Appetitive Desire in Plato and Aristotle.* New York: Oxford University Press, 2006.

Martin, Raymond, and John Barresi. *The Rise and Fall of Soul and Self: An Intellectual History of Personal Identity.* New York: Columbia University Press, 2006.

Mix, Lucas J. *Life Concepts from Aristotle to Darwin: On Vegetable Souls.* New York: Palgrave, 2018.

Shields, Christopher. *Aristotle.* New York: Routledge, 2007.

Walsh, Denis M. *Organisms, Agency, and Evolution.* Cambridge: Cambridge University Press, 2015.

CHAPTER 3

Vegetable Souls in the Middle Ages

Abstract Human experience as conscious organisms colors biology. We project this onto other organisms in ways that can hinder biology, particularly when we reify things like species identity, organismal individuality, and sex. Teleology was a key issue in biology from Aristotle to Darwin because of a difficult tension between the intentional teleology of humans and the mindless teleology of "lower" life. Interpretations of Aristotle shifted in parallel with a widening gap between human life concepts and vegetable life concepts, between psychology and physiology. Formal and final causes, central to Aristotelian soul-based biology, were increasingly associated with minds and, particularly for vegetable life, with the mind of God. This led to the coevolution of three concepts important for biology: nature, soul, and breath. Each one showed variation across time and place but addressed some aspect of natural teleology. Neoplatonic and Medieval biology moved toward a dualism that associated mind with active and masculine traits, body with passive and feminine traits. Souls still bridged the gap, but tensions increased leading toward more ideal psychology and more material physiology.

Keywords Dualism • Physiology • Psychology • Scala naturae • Teleology • Vegetable soul

L. J. Mix, *The End of Final Causes in Biology*, https://doi.org/10.1007/978-3-031-14017-4_3

Humans have a peculiar view of life, shaped by our experience as conscious, deliberative organisms. This perspective colors biology. At a surface level, we can ask how consciousness and reason arise in physiological context. At a deeper level, we can ask how our physiology and environment constrain the questions we ask and the answers we find satisfying. We cannot help projecting our own experience onto other organisms and other species (Mix 2009, pp. 58–69, 273–288).

Teleology was a key issue in biology throughout the period from Aristotle (384–322 BCE) to Francis Bacon (1561–1626 CE) precisely because of a difficult tension between the intentional teleology of humans and the mindless teleology of "lower" life. What kind of biological agents are appropriate when talking about plants? How can they have power, form, and function? Do they even come in discrete units?

Chapter 2 looked at the perennial questions of biology as they were framed by Aristotle, including the role teleology played in his thinking and the categories he established for talking about life. Over the next 2000 years, interpretations of Aristotle shifted in parallel with a widening gap between human life concepts and vegetable life concepts. There was not yet a divide between mind and nature, between the standard agency of humans and the "agency" of other biological agents. But they grew farther apart. Form and function, central to Aristotelian soul-based biology, were increasingly associated with minds and, particularly for vegetable life, with the mind of God. This chapter turns to that process, with an eye toward psychological life as it came unmoored from physiological context.

A more extensive treatment focused on vegetable souls and vegetable life concepts can be found in Mix (2018). Martin and Barresi (2006) and Goetz and Taliaferro (2011) provide histories of the soul more focused on the human aspects of life. Central to the debate throughout will be a question of whether, and to what extent, power, form, and function can be attributed to non-human life.

The Human Perspective

Humans, along with most large plants and animals, reproduce sexually. This leads to populations with (relatively) clear boundaries defined by reproductive isolation (i.e., the biological species concept). It results in a (relatively) clear distinction between individuals as diploid multi-cellular organisms, genetically distinct and separated from parents and offspring by a gamete stage. Mammals and flowering plants have morphologically distinct gametes and often dramatic sexual differentiation during development.

For centuries, these commonalities led biologists to reify species identity, organismal individuality, and sex in ways that can hinder our understanding of living beings. The previously unimagined breadth and diversity of the Bacterial and Archaebacterial Domains reveals a plenitude of species reproducing asexually while regularly engaging in horizontal gene transfer. Fungi display sexual reproduction with numerous mating types. Inclusive fitness and selfish gene theories have suggested—and observation established—that cooperation and competition occur at a variety of scales, even across scales, as when genetic and organismal interests collide in biological altruism. Even when considering more traditional plant and animal species, endosymbiosis and ring species violate classical assumptions. From the recognition of mutable species in the eighteenth century to the recognition of intra-genomic competition in the twentieth, the tale of modern biology can be told as a gradual deconstruction of historical categories, once thought essential and eternal.[1]

This creates tension between two perspectives. On the one hand, biologists want clear, distinct individual units (e.g., species, organisms, and genes). Such units reflect the distinctness of humanity (among species) and individual persons (among humans). They facilitate experimental isolation and mathematical description. On the other hand, biologists have always observed blurred lines between biological agents. In many ways, Platonism and Neoplatonism emphasized the continuity of particular individuals with ideal species and ideal species with divine action.

Modern thinkers default to a view of biology based on discrete autonomous units. Biological agents must be distinct individuals, with localized power, definite form, and intrinsic function. Standard agents must have a mind clearly distinguishable from its physical surroundings, capable of holding information and intention. Pre-Enlightenment thinkers considered a broader range of teleologies. Comparison of different biological agent concepts reveals a historical movement toward discrete units based in ideas about human spiritual and subjective life.

---

[1] See Godfrey-Smith (2009, 69–86) for many concrete examples.

THE NATURE OF LIFE

Plato and Aristotle both viewed biological agents as intermediaries between eternal ideal forms and changeable tangible particulars. They had the power (or participated in the power) to order the cosmos. Plato had built downward from a cosmic soul, while Aristotle built upward from elements, but both used souls as dynamic intermediates, form informing matter. Not everyone was comfortable with this compromise, however. Even in the Ancient Mediterranean, alternative approaches flourished. Most invoked the souls of Plato and Aristotle but reframed them in important ways. Epicureans, Stoics, and physicians leaned toward the physical. They spoke of animal souls and plant "natures," treating both as fully corporeal and clearly teleological. Plotinus and the Neoplatonists would later swing in the opposite direction, emphasizing the transcendent qualities of all living things.

The debate hangs on shifting definitions for three terms: nature, soul, and breath. The first attempts to identify regularity in the world. The latter two attempt to capture the dynamic continuity of living things. At one time, soul and breath were tightly coupled ideas about the power of biological agents, but that changed. Eventually breath and vegetable souls were associated with nature, while souls, especially human souls, were not.

Physis—Natura—Nature

By the early modern period, common perceptions of knowledge would be divided into three categories based on subject matter and methodology. Students of divinity discussed the properties of God, students of humanities discussed human properties, and students of natural sciences discussed the properties of everything else. These divisions were not in place, however, prior to the Renaissance. Augustine (354–430 CE) could easily ask about the nature of God and the divinity of insects.

The Greek word *physis* refers to the qualities of a thing, the properties associated with it. Latin scholars used the word *natura*. Pliny the Elder (23–79 CE) wrote a book called *Natural History*. He attempted a systematic and holistic approach to all forms of knowledge, but the term came to be associated with observational, descriptive, and often encyclopedic approaches to the world. Early modern biology is often presented as natural history prior to Darwin because it lacked a coherent theoretical underpinning. Thus, the famous quote attributed to Ernest Rutherford that biology is simply "stamp collecting" (Lewin 1982) or Michel Foucault's (1994, p. 269) assertion that *biology* per se only became possible with Georges Cuvier (1769–1832).[2]

More theoretical and mathematical approaches came to be called natural philosophy, as in the title to Isaac Newton's 1687 book *Mathematical Principles of Natural Philosophy*. Nonetheless, the term readily incorporated immaterial agents well into the nineteenth century, as in the *Spirit* of German *Naturphilosophie*.

Natural science in the modern sense arose during the Enlightenment, and this was the version of "nature" which most often informs current discussions. Over time, philosophers began to speak of nature as a distinct realm of passive, tangible, extended objects. Influential proponents include Francis Bacon, René Descartes, Robert Boyle, and John Stuart Mill.[3] There has never been a time when a clear and consistent line could be drawn between natural and non-natural explanation. Rather, there has always been ambiguity around the relationship between consciousness and materiality with "nature" referring, ambiguously and alternatively, to both the fundamental essence of a thing (known to reason) and the material regularity of a thing (known by observation).

Enlightenment thinkers explicitly excluded formal and final causes from their mechanical and material world, the "nature" of natural science, in order to create a coherent, consensus epistemology. The "naturalization of biology," then, refers to attempts to ground biological agents, form, and function in the material world after they had been explicitly excluded.

---

[2] Though popular references, both attributions can be misleading. It is unclear whether Rutherford ever made this remark and Foucault was emphasizing an epistemic shift related to new conceptions of "man" as subjective observer, rather than life as the object of observation. For a closer look, see Foucault (1994, especially pp. 207 and 246) and Gutting and Oksala (2021).

[3] Francis Bacon (1561–1626) distinguished physics from metaphysics in his *New Organon, or True Directions Concerning the Interpretation of Nature* (1620). Note the use of the term "organon," following the Medieval compendium of Aristotle's works on logic. It suggests that knowledge, like an organism, has both structure and purpose. René Descartes (1596–1650) distinguished between one realm of extended things and one of thinking things or minds. Robert Boyle (1627–1691) pioneered experimental reasoning, promoting mechanical philosophy and empirical methodology. John Stuart Mill (1806–1873) identified a popular ambivalence about the inclusion of humans in nature in his 1874 essay on "On Nature." For more on the teleology and the origins of modern science, see Osler (2010).

To understand why and how the modern naturalization works, one must understand the de-naturalization of souls that preceded it.

Physis comes into English as physics and physiology, both of which attempt to explain observed regularities in nature. The first applies generally to extended objects; the latter applies more specifically to the regularities of life. Aristotle made a point of calling *psyche* the *physis* of organisms. And so, psychology—the study of souls—was once the study of the nature of living things. By the end of the Enlightenment, however, physiology and psychology would become radically distinct pursuits.

Psyche/Pneuma—Anima/Spiritus—Soul/Breath

The terms for soul, *psyche* in Greek and *anima* in Latin, could refer to a host of biological agent concepts, from Platonic immaterial and eternal motivators to Aristotelian material processes and Epicurean rarified material bodies. They consistently referred to life, including humans but also—almost universally—the life of non-human animals. Frequently, though not universally, soul language was extended to plants as well—to every living thing.[4]

One property was consistently associated with souls: breath. The terms *pneuma* (Greek) and *spiritus* (Latin) described something at the boundaries of non-biological nature. On the one hand, it could be viewed as a special kind of air, refined by and empowering living things: physical breath and breathing. It had this quasi-material connotation from Aristotle to Descartes. On the other hand, it could be explicitly transcendent, as in the Holy Spirit of Christianity, literally the breath of God, or the Spirit of the German Idealists, directing history. Prior to the Enlightenment, both soul and breath should be interpreted with caution, recognizing their duality, which—often quite intentionally—bridged the gap between physical and mental concepts of life. Breath conveyed or described the power, form, and function that make life life-like.

A soul was identified with a living, breathing thing. It was the biological agent of choice for Christian and Islamic cultures around the Mediterranean in Antiquity and the Middle Ages. For some thinkers, this meant a discrete autonomous entity with power, form, and function of its own. For others it referred to participation in a larger life, whose power, form, and

---

[4] Prominent proponents of vegetable souls include Plato, Aristotle, Lucretius, Pliny, Origen, Plotinus, IBN Sînâ, Maimonides, Aquinas, Gassendi, Leibniz, and Hegel.

function moved the cosmos. In either case, it referred broadly to all living things, not just humans.

Between the fifth century BCE and the sixteenth century CE, distinct theories of mental life developed that distinguished the nature of minds (Greek *nous*, Latin *mens*) from the nature of vegetables. They addressed spirituality (e.g., transcendence, participation in the divine), subjectivity (e.g., interior life, consciousness), and reason (e.g., intellect, cognition). This reified the distinction between human and non-human biological agents.

The category of "vegetable" developed in this period to describe living things with nutrition, development, and reproduction, but none of the mental activities of humans. It does not refer to the kingdom Plantae, but to all "lower" organisms.[5] Meanwhile, the category of "animal" existed awkwardly in between vegetable and human, with some of the properties of mental life, but not all. Philosophers continued to speak of all humans as animals and all animals as vegetables in a technical sense, but the terms came to be exclusive in casual usage.

Aristotle had proposed that animals have sensation and willed motion, but both were debated at great length. What do those activities entail? And how do we know who has them? In *Life Concepts*, I argue that the traditional definitions map poorly onto modern biology (Mix 2018, pp. 239–248). Physically, all organisms react to their environment and change it in turn. Metaphysically, empirical science has no way to adjudicate a "back-end" of subjective interiority nor an ability to speak of its connection with "front end" physiology. Biologists cannot determine whether all humans have consciousness or cognition in any transcendent way, much less spiritual or intellectual contemplation. (Perhaps even a bat does not know what it is like to be a bat.) The things we can measure—stimulus and response—appear universal across known life.

Ancient, Medieval, and Renaissance thinkers struggled with the same issues. The evolution of nature, soul, and breath concepts reflects a changing intellectual environment as biologists, philosophers, and theologians grappled with the implications of different theories. To shoehorn any of them into an Enlightenment or modern dualism (body vs. mind, objective vs. subjective, scientific vs. theological) would be to obscure exactly the work they were intended to do—to bridge human and non-human life.

[5] Mix (2018, pp. 213–224) traces the evolution of vegetable life into kingdom Plantae and other taxonomic categories.

Teleology has always—and quite clearly—applied to human mental life. It describes intention associated with standard agency. Alison imagines two futures, prefers one, and has the ability to act on that preference. When she writes a computer program, she has the power to run it, she informs it (makes it what it is), and directs it (gives it a function).

The open question is whether we can apply teleology more broadly. Do human bodies have power, form, and function when no human mind is involved? And can these same properties be applied more broadly to animal and vegetable lives? In Late Antiquity, most thinkers took for granted animal ensoulment, but debated whether plants required such psychic explanation or could be understood with simpler, abiological physical accounts.

MATERIAL SOULS

Following Aristotle, several schools of thought attempted a fully physical biology (von Staden 2000; Gundert 2000; Mix 2018, pp. 68–71). Epicureans spoke of soul seeds, a special kind of atom with inherent motion that could motivate biological power. To explain nutrition, development, and inheritance, they appealed to souls that were corporeal: composed of atoms, both soul seeds and elements, and taking up space. The rarified qualities of these souls meant they could permeate and empower the grosser, tangible bodies of plants and animals. One of the most famous Epicureans, Titus Lucretius Carus (first century BCE) went further to invoke two parallel biological agents in humans, the feminine *anima* for vegetable activities and a masculine *animus* for rational activities.[6]

Stoics and many early physicians developed *pneumatic* theories, attributing power, form, and function to refined air running through the body. Erasistratus of Ceos suggested that the heart refines common air into life-breath (*pneuma zootikon*) which circulates with the blood and mediates vegetable functions. This breath was further refined in the brain to form the soul-breath (*pneuma psychikon*), which manages animal functions. In the same vein, Stoics referred to a nature-breath (*pneuma physikon*) and a soul-breath (*pneuma psychikon*), adding a ruling faculty for humans (*hegemonikon*). Curiously, they placed a biological ruling faculty in the heart, naming it the hottest part of the body, and a greater ruling faculty in the

---

[6] The contemporary Christian theologian Tertullian (ca. 155–ca. 220) also made this distinction.

Sun to rule the cosmos. These theories display a tendency to associate vegetable life with a physis or nature as distinct from the soul of animals. Nonetheless, the lines had not yet been clearly drawn. Galen of Pergamon (129–c. 200 CE), with some humor, confessed to using the terms flexibly: plant "souls" with Platonists and plant "natures" with physicians (*On the Natural Faculties* 1.1, Galen 1963).

These perspectives highlight an important trend toward dualism, associating the body with passive and feminine traits, mind with active and masculine traits. Passed on through Neoplatonism and theology, they would be influential in Medieval biology.

The Ladder of Nature

Souls were increasingly conceived along Platonic lines with power, form, and function flowing outward or downward from God. Plotinus (204/5–270 CE), the founder of Neoplatonism, described a divine unity, the One, which eternally moved, informed, and directed the cosmos through a series of stages or "emanations." First came a cosmic Intellect, then a cosmic Soul, and onward through intellects and souls and eventually material bodies.

The progression of decreasing freedom and dignity came to be known as the *Scala Naturae* or Ladder of Nature. It was reappropriated and reformatted in numerous ways, but within the biological world several features remained consistent. Lower life was always directed by and served the purposes of higher life. Rational life was higher than animal life, which, in turn, was higher than vegetable life. And all life was higher than rocks and other non-life. Across the scale, higher life was associated with masculinity, lower life with femininity. Thus, the male parent planted a seed (with necessary power, form, and function) within a feminine ground, the female parent (contributing matter only).[7]

Neoplatonic thinking took the idea of form informing matter in a new direction, more in line with Plato than with Aristotle, but containing important new features. Plato's rational soul had been a way to understand humans interpreting the order of the cosmos. Plotinus' emanations were a comprehensive theory about the nature of everything. In this model, the cosmos itself is a biological agent. Individual souls and

[7] Miller (2002) explores this idea as it enters modern biology through German Idealism.

individual organisms—and everything else—are organs in one, universal organism. Their ends were subsidiary organic functions, derived from the ends of the cosmos. Abiological entities were the exception, rather than the rule.

Plotinus' theory began in a very mental way, with a cosmic Intellect. It presented an explicitly organismal metaphor, wherein the power, form, and function of the cosmos could be found in cosmic organs of thought and purpose. Just as the heart derived its final cause (pumping blood) from the perpetuation of the organism, so every object in the universe derived its final cause from the cosmic Soul and, ultimately, from the cosmic Intellect. These ideas would prove tremendously appealing to Jewish, Christian, and Muslim theologians attempting to reconcile scripture with natural history throughout the Middle Ages. They would come to associate the One with the God of Abraham and final causes with ideas in the mind of God.

Physical and Spiritual Life

The teleology of Neoplatonism may seem a far cry from the practical materialism of Aristotle and the Ancient physicians, but it served an important purpose for Medieval biologists. It allowed them to talk about power, form, and function in vegetables without resorting to the, by then, ridiculous idea of vegetable minds. By outsourcing agency, identity, and intent to God, they could move forward with the more mundane concerns of plant physiology and the still difficult problems of nutrition, development, and growth. But how did these concerns get so far separated from mental life? That story begins with scriptural accounts of life and the foundations of monotheistic natural theology.

Hebrew Scriptures, that is, the Tanakh or Old Testament, had their own biological agent concepts (Mix 2018, pp. 80–83; Brown 2014; Cooper 1989). Vegetable life was treated as background, provisions, and furnishings for more active animal life. Agency, both human and animal, arose as breath stirring up the dust and mud of the earth. Genesis, the first book of the Tanakh, describes living beings dynamically. The movement of *ruach* in the dirt creates *nephesh chay*. Later translators render *ruach* as *pneuma-spiritus-breath* and *nephesh* as *psyche-anima-soul*, though it might also be translated as living, breathing thing. In Genesis 2:7, God's breath transforms the dirt into Adam, the first human, but the same word (*nephesh*) appears when other animals are created in Genesis 1. This emphasis on God as ultimate efficient cause aligned well with both Aristotelian ultimate causes and Neoplatonic emanations.

Early Hebrew life concepts suggest that human and animal agents are fully mortal, but a new concept arose just before the common era: life after death. Later Hebrew Scriptures (e.g., Isaiah 26:19) describe human resurrection. It remains unclear whether the original readers interpreted this as eternal life or simply a second round, but in either case this second life was motivated by divine breath and, thus, divine agency. Human life became spiritual life.[8]

The Christian New Testament provides extensive commentary on spiritual life. It is associated with word (*logos*), and hence breath, and the fundamental order of the cosmos (e.g., John 1). Spiritual life comes to be associated explicitly with humans and it can be difficult to tell how terms like *pneuma* and *psyche* should be interpreted. The liminal quality of breath is clear from expressions for death such as *ekpneo* (literally to breathe out) and *ekpsycho* (literally to yield up the soul), translated into English as "gave up the ghost" or "yielded up his spirit" as well as simply "died."[9] The human soul, separated from the power, form, and function of divine agency, dies. It was not a subsistent autonomous agent, but a dependent and relational one.

Jewish and Christian theologians early in the common era sought to integrate scriptural life concepts with the biology of physicians and natural historians. They found easy correspondences between a God who creates by breath (and spoken word), the ultimate cause of Aristotle, and the One of Plotinus. Jewish theologian Philo of Alexandria (ca. 20 BCE–ca. 40 CE) had a particularly strong influence through his dual creation theory (Mix 2018, pp. 86–88). He divided creation into two distinct acts.[10] The invisible creation (Genesis 1:1–3) refers to God making intelligible forms in eternity. The visible creation (Genesis 1:4–2:3) describes how the forms became manifest in time and matter. This dualism helped to reify Aristotle's formal and final causes. Through the Middle Ages, the formal causes developed into immutable (eternal) species while final causes became

---

[8] "Thus says the Lord God to these bones: I will cause breath to enter you, and you shall live. I will lay sinews on you, and will cause flesh to come upon you, and cover you with skin, and put breath in you, and you shall live; and you shall know that I am the Lord.'" Ezekiel 37:5–6, New Revised Standard Version.

[9] *Ekpneo* in Mark 15:37 and 15:39 as well as Luke 23:46; *Ekpsycho* in Acts 5:5, 5:10, and 12:23.

[10] *On the Account of the World's Creation Given by Moses*, in Philo (1949).

God's intentions for living beings. Later theologians divided creation in different ways but retained the distinction between an eternal formal act and a temporal physical process.[11] This had several curious consequences for natural philosophy. For example, some theologians placed human souls in eternity and vegetable souls in time. Others, noting no mention of planets in Genesis 2, would distinguish the motive principles of planets (in eternity) from the motive principles of terrestrial animals (in time).

Philo went on to explain Genesis 2 as a way of understanding human composition.[12] Echoing the pneumatic theory of earlier thinkers, he wrote of a material body composed of earth and a material soul composed of rarified air. He further interpreted Adam as the human intellect and Eve as the human body, replaying in microcosm the two stages of creation. First God made human minds and then, from them, human bodies. This fit well with a hierarchical understanding in which "lower" life was viewed teleologically as instrumental to "higher" life—body *for* mind and woman *for* man. This, too, would be incorporated into Medieval dichotomies.

Despite the oppositional quality of Philo's pairs—active/passive, soul/body, male/female, spiritual/physical—there remained a Platonic or Neoplatonic continuity. No created thing had power, form, or function in and of itself. They had these qualities due to participation in the greater cosmic life. All of creation was, ultimately, dependent on divine teleology. All agents, including biological agents in humans, animals, and plants, were active with respect to abiological elements and "lower" forms of life. To be active, to be an agent, was to be superior. And yet, all agency was only possible because of passive participation in "higher" forms of life. Divine agency moved all lesser agents. True dualism would require a much more radical shift.[13]

---

[11] Augustine distinguishes an invisible creation in Genesis 1 from a visible one in Genesis 2–3. He deals with this in *Confessions* chapters 12 and 13, *City of God* chapter 11, and most fully in his latter *On the Literal Interpretation of Genesis*. Aquinas followed his lead in *Summa Theologiae* 1.44–74.

[12] *Allegorical Interpretation of Genesis 2 and 3*, in Philo (1961).

[13] A similar continuity can be seen in the New Testament in I Corinthians 15:44, which prefers the resurrected spirit-oriented body (*soma pneumatikon*) to the pre-resurrection soul-oriented body (*soma psychikon*). The problematic translation in the New Revised Standard Version of the Bible as a spiritual body and a physical body reflects a modern dualism and not a traditional split. Worse still, it has often been interpreted with the soul on the spiritual side, when the text places it on the physical side.

METABOLISM AND CONSCIOUSNESS

Augustine of Hippo (354–430) may be viewed as a hinge between Mediterranean Antiquity and Medieval Christendom in Western Europe. He developed a new synthesis of Plato, Aristotle, and Hebrew perspectives in which the dual creation played a critical role (Mix 2018, pp. 103–115; Goetz and Taliaferro, pp. 35–44). He set forth a system for understanding plant and animal natures dependent on eternal kinds. These kinds were generated in the invisible creation and, through seeds, communicated power, form, and function from one generation to the next. Drawing on the Hebrew word for kind (*miyn*) and the Septuagint Greek translation (*genos*), he used the Latin term *genus*, still in use today. Though he occasionally referred to vegetable souls, he usually reserved the term "soul" for animals that breathe and move and sense their environment.

Augustine's categories reveal a key innovation. Like Aristotle, he used sensation to differentiate animals from plants, but he did so in a distinctly Platonic way. He spoke of souls as immaterial substances that grant power to living things. Nutrition, sensation, and so on shifted from material processes to biological faculties. Souls became distinct in a new way, disembodied biological agents acting through instrumental bodies.[14] The form (formal cause) and function (final cause) of living things moved into an immaterial world of minds.

The soul itself must be immaterial because it can hold immaterial images. The more rarified elements, fire and air, share an affinity with spiritual life that allows them to mediate between the grosser (water and earth based) material body and the immaterial soul. Even brute animals can remember and anticipate and, thus, must have some form of immaterial representation. This produced, perhaps for the first time, a concept of subjective or interior life, distinct from objective physical processes (Matthews 2000; Martin and Barresi 2006, pp. 69–72; Goetz and Taliaferro 2011, pp. 32–47). Intellect, which had been a holistic process of living things receiving forms, became a faculty whereby an immaterial soul perceives (or participates in) divine thought—without appeal to the senses.

---

[14] Mix (2018, pp. 118–119) looks more closely at the reification of the soul in Late Antiquity. "Thus, the human soul became an incorporeal cause of corporeal activity. I will refer to such souls as *subsistent* because they support themselves without either material cause (parts of which they are made) or matter (stuff which they shape). Subsistent souls inform matter but are more than the form of matter. They provide the potential or power for vital activities and exist prior to them."

Just as sensation reflected the impact of the physical world on a soul, so locomotion became the impact of a soul on the physical world. The internal could move the external. In humans, this faculty of motion was associated with both intellect and will. It might, somewhat loosely, be called informed action. This provided the foundation for causal efficacy in the standard story of agency—minds causing changes in the physical world. Augustine and contemporaries believed that the internal/external interface could be crossed in both directions for all animals, albeit in a freer and more sophisticated way for humans.

Augustine set up the divide between internal and external life, presaging a subjective/objective distinction for experience and mechanical/willed distinction for action. Throughout the Middle Ages, biologists and philosophers would continue to recognize these as aspects of a continuum. All matter participated in the eternal forms, and those forms were universally real yet apprehended internally. And yet, the pieces were now in place for a genuine mind-body dualism, rationalism and empiricism, and the development of modern science.

Consciousness became a distinctly animal phenomenon. Plants lacked the interiority necessary for immediate or intrinsic function. Thus, the application of Aristotelian formal and final causes to plants required remote or extrinsic function—someone's intention for them. The obvious choice was God, whose mind could hold form and function enough from the entire world. Formal and final causes became distal (possibly ultimate) efficient causes.

In earlier paradigms, biological agency reflected the continuity of vegetable life with cosmic life. By the end of the Middle Ages, form and function would be excluded from vegetable life and pushed upward into the mind of God, leaving vegetable souls either clearly theological or ambiguously defined. If the human body was an instrument of the human soul, of what then were animal and vegetable bodies instruments?

THINGS FALL APART

For centuries, biologists struggled to make sense of vegetable and animal life. They sought to maintain the traditional continuity of all living things using Aristotelian souls and Neoplatonic hierarchies while developing new theories of spiritual life and consciousness that did justice to human experience. The nature of souls and life were popular topics for the Islamic *Falsafah* and Christian Scholastics (Mix 2018, pp. 119–140). Biological agent concepts—often called vegetable souls—proliferated, mutated, and competed around the Mediterranean and Western Europe. Debates raged over the singleness or multiplicity of souls in a single human, the materiality of souls, the presence of souls in plants and animals, and more broadly the interpretation of Aristotelian causes. Prominent theologian/philosophers tackling these questions included Muslim AL-Fârâbî (870–950), IBN Sînâ (or Avicenna, ca. 970–1037), AL-Ghazâlî (ca. 1058–1111), and IBN Rushd (or Averroes, 1126–1198); Jewish Moses ben Maimon (or Maimonides, 1138–1204); and Christian Albert the Great (1200–1280), Thomas Aquinas (1225–1274), and William of Ockham (ca. 1287–1347).

A wide variety of biological agent concepts were considered in this period, but two important selection events can be traced to Albert the Great who, through his student, Thomas Aquinas, had a tremendous impact on Western biology. Albert identified humans with their intellects. Following Plato and Augustine in speaking of two creations, he named humans the only embodied spiritual creatures. He still subscribed to a theory of divine agency moving the cosmos but made humans into a unique kind of agent. Human souls were embodied, but capable of existing without bodies. In this, he explicitly rejected Aristotelian hylomorphism and entelechy.

Aristotle had described souls as a union of matter (*hyle*) and form (*morphe*). They exist dynamically in a state of fulfillment (*entelechia*) because nutrition and other processes continuously sustain them. Albert asserted that human souls were different, distinct from both matter and activity and unique in creation. God creates each individually. Intellect (and by extension sensation and nutrition) should not be understood as an in-fulfillment activity, but as an *in potentio* faculty.

At the same time, Albert embraced natural philosophy and natural theology (Führer 2022). He spoke of the value of matter and the value of physical observation, composing extensive descriptions of both plants and animals. He was happy to describe non-human life and souls as hylomorphic, mortal, and contingent along very Aristotelian lines. In his *Summa Theologiae*, he stated that he follows Plato on the eternal immateriality of the (human) soul, but Aristotle on the dynamic physicality of the soul (specifically the vegetable soul or vegetable faculties of the human soul). Despite this move toward dualism, he still asserted a continuum. Plants are more living than rocks, but less so than humans. He articulated the Ladder of Nature clearly, drawing parallels with the nested hierarchies of feudal society.

Albert's position became systematized in the two distinct ladders of Thomas Aquinas. They may, with only slight anachronism, be considered supernatural and natural.[15] The invisible creation, including causes and ideas in the mind of God, extends downward through ranks of angels, stars, and planets to human souls—all possessing the faculty of intellect. Of these, only human souls were tangibly embodied. The visible creation, including human bodies, extends from humanity downward through animals, plants, and the abiological objects of the sublunary world. Thus, the human agent and the biological agents of plants, the human soul and the vegetable soul, became ontologically distinct. One was immortal and transcendent, the other mortal and necessarily embodied.

Humanity's unique position on both ladders proved essential to Renaissance and Enlightenment views of knowledge. Participating in the invisible creation, humans could perceive thoughts in the mind of God, including formal and final causes. Participating in the visible creation, humans could perceive individual examples, reason about particular instances, and apply eternal knowledge to temporal problems. The first involved intellect, a faculty of the immaterial soul or mind. The second involved sensation and willed motion, faculties of the material body (and, at the time, the animal soul). Biologists could be said to "know" things about life because, through their intellect, they perceived the essences and ends present in the mind of God. They could know things about particular creatures because they interacted with them physically.[16]

* * *

The unified biology of Aristotle—built around souls as a confluence of efficient, formal, and final causes, form informing matter in action and in fulfillment—gave way to two distinct approaches to life—a physiology of

---

[15] One of the themes of this book is the evolving character of "nature" over the centuries. By the time of Aquinas, we can distinguish between the material visible creation and the immaterial invisible creation. For Aquinas, "nature" and natural law still referred to the character of things and not to the physical universe, but the idea of a physical universe, with regular laws (God's ordered power) and amenable to observation, was in place. See Osler (2004) for an extended discussion of scholastic debates on God's absolute versus ordered power and their contributions to the foundations of modern science.

[16] Among spiritual beings, humans alone need to do science; higher intellects already know the truth of eternal forms. Among physical beings, humans alone can do science; lower forms of life lack the ability.

bodies and a psychology of minds. This allowed a technical language to develop for each, leading to reflection on the details of vegetable and mental life. In the first, biological agents express the power, form, and function of God. In the second, rational agents possess intrinsic power, form, and function which may or may not align with the will of God. (There was some debate about whether non-human intellects could also stray.)

The division of rational and biological agency created a chasm between human and vegetable life concepts. Animals remained awkwardly in the middle, sharing some but not all the mental features of humanity. Ancient and Medieval debates largely focused on the plant/animal division, asking whether "souls" could be applied to both. Enlightenment debates would shift the focus to the animal/human distinction, but the same issues remained important. In what ways does non-human life share in the interiority and spirituality attributed to humans? How is mental life grounded in physicality, physiology, and animality?

The next chapter will turn to Enlightenment philosophers, led by Bacon and Descartes, who tried to set boundaries around nature to facilitate good observation, critical reasoning, and consensus building in physics and physiology. By excluding minds from natural science—along with agency, form, and function—they created new tools for reasoning about the world.

The passive, mechanical nature they defined would enable a powerful new physiology to develop and clear the way for nominalist and evolutionary theories about life. It would place formal and final causes out of bounds, but still require theories of power, form, and function. Vegetable souls still could not hold intrinsic form and function, but neither could they be used as pointers to the mind of God. Thus, vegetable souls became nonsensical. New biological agents would have to be developed. Physiology and psychology had drifted apart, making it difficult to speak—or even think—of humans as part of nature. These problems would be addressed, if never completely solved, by natural selection, genetics, and the Modern Synthesis, but the intervening centuries provided a unique environment for innovation.

References

Brown, William P. "Life." In *Oxford Encyclopedia of the Bible and Ethics*, edited by Robert L. Brawley. New York: Oxford University Press, 2014.

Cooper, John W. *Body, Soul, and Life Everlasting*. Grand Rapids, MI: Eerdmans, 1989.

Foucault, Michel. *The Order of Things: An Archaeology of the Human Sciences.* New York: Vintage, 1994.

Führer, Markus. "Albert the Great." In *Stanford Encyclopedia of Philosophy*, Summer 2022 ed. Stanford University, 1997–. https://plato.stanford.edu/archives/fall2017/entries/albert-great/.

Galen. *On the Natural Faculties.* Loeb Classical Library. Cambridge, MA: Harvard University Press, 1963.

Godfrey-Smith, Peter. *Darwinian Populations and Natural Selection.* New York: Oxford University Press, 2009.

Goetz, Stewart, and Charles Taliaferro. *A Brief History of the Soul.* Malden, MA: Wiley-Blackwell, 2011.

Gundert, Beate. "Soma and Psyche in Hippocratic Medicine." In *Psyche and Soma: Physicians and Metaphysicians on the Mind-Body Problem from Antiquity to Enlightenment*, edited by John P. Wright and Paul Potter, 13–35. Oxford: Clarendon, 2000.

Gutting, Gary, and Johanna Oksala, "Michel Foucault." In *Stanford Encyclopedia of Philosophy*, Summer 2021 ed. Stanford University, 1997–. https://plato.stanford.edu/archives/sum2021/entries/foucault/.

Lewin, Roger. "Biology is Not Postage Stamp Collecting." *Science* 216 (1982), no. 4547: 718–720.

Martin, Raymond, and John Barresi. *The Rise and Fall of Soul and Self: An Intellectual History of Personal Identity.* New York: Columbia University Press, 2006.

Matthews, Gareth. "Internalist Reasoning in Augustine for Mind-Body Dualism." In *Psyche and Soma: Physicians and Metaphysicians on the Mind-Body Problem from Antiquity to Enlightenment*, edited by John P. Wright and Paul Potter, 133–145. Oxford: Clarendon, 2000.

Miller, Elaine P. *The Vegetative Soul: From Philosophy of Nature to Subjectivity in the Feminine.* Albany, NY: State University of New York Press, 2002.

Mix, Lucas J. *Life in Space: Astrobiology for Everyone.* Cambridge, MA: Harvard University Press, 2009.

Mix, Lucas J. *Life Concepts from Aristotle to Darwin: On Vegetable Souls.* New York: Palgrave, 2018.

Osler, Margaret J. *Divine Will and the Mechanical Philosophy.* New York: Cambridge University Press, 2004

Osler, Margaret J. *Reconfiguring the World: Nature, God, and Human Understanding from the Middle Ages to Early Modern Europe.* Baltimore: Johns Hopkins University Press, 2010.

Philo. *Philo*, 2 vols. Translated by F.H. Colson and G.H. Whitaker. *Loeb Classical Library.* Cambridge, MA: Harvard University Press, 1949.

Philo. *Philo Supplement I: Questions and Answers on Genesis.* Translated by R. Marcus. *Loeb Classical Library.* Cambridge, MA: Harvard University Press, 1961.

von Staden, Heinrich. "Body, Soul, and Nerves: Epicurus, Herophilus, Erasistratus, the Stoics, and Galen." In *Psyche and Soma: Physicians and Metaphysicians on the Mind-Body Problem from Antiquity to Enlightenment,* edited by John P. Wright and Paul Potter, 79–116. Oxford: Clarendon, 2000.

CHAPTER 4

Mechanical Organisms in the Enlightenment

Abstract Enlightenment Biologists looked for new ways to think about nature. Proponents of empirical and mechanical philosophy sought a physiology, a physics of life, free from the problems of interiority, intellect, and spirituality. Natural science in the seventeenth century and natural selection in the nineteenth century dramatically changed the population of biological "agents," the basic units of biology, loci of power, form, and function. Bacon and Descartes rejected the "Aristotelian" formal and final causes central to earlier theories of vegetable souls. This established a fully natural biology and physiology but shifted teleological questions backward to Creation and upward to the mind of God, supporting appeals to intelligent design. Kant and Darwin proposed new ways of thinking about power, form, and function grounded in an etiological recursion and a chain of efficient causes. Darwin's new theory of evolution by natural selection avoided appeals to transcendent, progressive, and mental teleologies and focused on change in populations rather than individuals. It opened the door for natural teleology.

Keywords Darwin • Evolution • Mechanical philosophy • Natural selection • Teleology

L. J. Mix, *The End of Final Causes in Biology*,
https://doi.org/10.1007/978-3-031-14017-4_4

Where do power, form, and function come from? Living things seem to be uniquely equipped to change their environment and to maintain themselves while being changed by it. Lineages of genes, cells, organisms, and species persist through time in a way that differentiates them from their surroundings. Modern biologists take such units for granted, but they have a history that shapes the way they are used.

Aristotle explained the dynamic consistency of life by appealing to vegetable souls, a confluence of efficient, formal, and final causes engaged in nutrition, growth, and development. These souls "acted" on the world despite lacking minds, intentions, or the kind of agency attributed specifically to humans after the Enlightenment.

As centuries passed, many biologists continued to use the language of formal and final causes, but their understanding of the terms changed. As physiological and psychological perspectives diverged, formal and final causes were increasingly viewed as mental rather than physical and, thus, beyond the reach of "lower" vegetative life forms. Medieval thinkers placed final and formal causes in the mind of God (Chap. 3; Osler 2004, p. 162). This provided a transcendent repository for form and function that could be accessed by vegetable souls without requiring those souls to have a mind of their own.

Many Ancient and Medieval thinkers viewed the cosmos as a single, biological entity, making this displacement less dramatic than it may sound to modern readers (Ruse 2010). Individual living beings could be organs of a greater organism. The final cause came to be interpreted as transcendent ends, outside the physical world.

During the Enlightenment, natural historians and natural philosophers—the forerunners of modern biology—looked for new ways to think about nature. Proponents of empirical and mechanical philosophies sought a physiology, a physics of life, hermeneutically sealed from the difficult problems of interiority (i.e., consciousness and subjectivity), intellect (i.e., a faculty for perceiving and knowing truth), and spirituality (i.e., divine or resurrection life). They sought to once again locate biological agency in the physical world, to return the proper ends of organisms to the organisms themselves or some aspect of them.

The rise of natural science in the seventeenth century and natural selection in the nineteenth dramatically changed the basic units of biology—loci of power, form, and function or biological "agents." This began with the dismantling of vegetable souls and ended with the rise of genes. In between, there was a proliferation of theories.

THE EFFECTS OF NATURAL SCIENCE

The Scientific Revolution put stress on the concept of vegetable souls. In the seventeenth century, empirical and mechanical philosophers responded to Medieval biology by rejecting final and formal causes. To be precise, they removed them from the world known to the senses. Francis Bacon (1561–1626 CE) and René Descartes (1596–1650) devised new ways of thinking. By excluding minds, they redefined "nature" as matter in motion. They argued that Aristotle's material and efficient causes were sufficient for natural science—and usually all humans have access to. This shift destabilized the Medieval framework for biology. Bacon and Descartes' followers asked which, if any, version of vegetable souls could survive in the new conceptual environment.

Natural science set the mind of God off-limits. Bacon (2000) compared final causes to consecrated virgins: beautiful and worthy, yet unapproachable. We should focus on their servants, the efficient causes. One acts through the other. God knows the true form (formal cause) and function (final cause) of all things. We can only know what they are made of (material cause) and how they move (efficient cause). Bacon called the first two metaphysics and the latter two physics. Descartes, meanwhile, argued that "We shall not select any reasons about natural things concerning the end that God or nature intended for himself in creating these things: because we ought not be so arrogant as to think that we are participants in his plans" (quoted in Osler 2004, 213). Transcendent ends exceed our reach. Bacon and Descartes focused attention on composition and movement. They initiated a "mechanical" model of the universe (Osler 2004, 2010; Mix 2018, pp. 143–174).

Mechanism caught on quickly in physics, but not biology. Power, form, and function could not be reconciled with a mechanical world. Medieval biologists had reimagined Aristotle's biological agent concept, the soul— changing it from a process to an ideal entity responsible for the process, an ultimate cause or at least a very special kind of cause now associated with either human or divine intention. This strange entity, now thought of as an "agent" with transcendent causal "power" was incompatible with vegetable life, potentially all non-human life. Along with form and function, they required an intentional mind.

To be clear, the mechanists recognized that the vegetable soul of Aristotle and many later interpreters was not such a thing. But, in reifying the rational soul of humans (now immortal, transcendent, and intellectual) and mentalizing form and function, they lost sight of how vegetable souls might be useful explanations. They also recognized that power, form, and function were key aspects of non-human biology. Plants and animals appeared to be more than simply the mechanical nature of physics, but less than the full-fledged (standard story) "agents" called human souls.

Various awkward patches were considered. Bacon added quasi-mechanical *spirits* to power the machines. Leibniz invoked a miraculous pre-established harmony between mechanical bodies and immaterial *monads*. These biological actors took the place of animal souls—allowing for sensation and willed motion—but continued to sound odd in the vegetable context. More problematically, they still stretched the boundaries of nature in controversial ways. Natural scientists struggled with vegetable organization and behavior.

Cartesian Biology

Descartes bit the bullet.[1] Everything in the physical world, he said, including plants and non-human animals, operates mechanically and deterministically. Cartesian matter is passive, inert. It moves deterministically as a result of God's power in the distant past and human power in the present. This may sound appealing to modern thinkers, but it contained a crucial flaw. Descartes still believed that plants were organized. Without a natural locus for information and purpose, organization could not arise within the machine. Biological organization must have been built in by a designer.

Mechanism fits easily with intelligent design arguments. Gassendi, Boyle, and Newton all saw a divine hand in the operation of the universe (Osler 2004, pp. 56–59). Mechanists accused of atheism (for removing God's power from natural explanations) could point to the machine metaphor and say that machines reveal something about their maker. Gassendi, in particular, used final causes as providential efficient causes (p. 161). Descartes critiqued this as God inelegantly tweaking creation—tinkering with a machine that should already be perfect. In the following centuries, his view came to dominate. Final causes came to be understood as direct divine intervention *contrary to* normal physical causation.

Descartes' dualism also pushed human intentions out of the natural world—along with consciousness, will, and reason. They could not arise

---

[1] Mix 2018, pp. 151–154; see Descartes' *Discourse on Method*, part 5.

within nature, only act upon it mysteriously, perhaps miraculously, at the pineal gland (Lokhorst 2017). All power, form, and function that did not arise from human action (artifice) must have been present from the beginning of time.

Baconian Biology

Bacon tried to bridge the gap using a broader definition of matter (Klein 2016; Rees 1980). Echoing the Epicureans, he suggested biological actors as intangible bodies. In addition to inert *tangible matter*, Bacon invoked active *pneumatic matter*. (This links him to the "chemical" as well as the "mechanical" philosophy.) Weightless and invisible, pneumatic matter nonetheless imparts motion (like soul seeds). Pneumatic bodies, or spirits, bind tangible matter into tangible bodies. Vegetable spirits, composed of rarified pneumatic matter, give power, form, and function to plants. Bacon did not specify how spirits form, how they interact with less rarified matter, or how they hold information and purpose.

Kantian Biology

Immanuel Kant (1724–1804) provided an epistemological compromise (Mix 2018, pp. 163–164).[2] He eschewed the ontological dualism of Descartes (mind and matter) and the "natural" dualism of Bacon (tangible and pneumatic matter). With Augustine, he recognized a difference between external and internal worlds. The realm of appearances (*phenomena*) reflects our internal sense of the external world, our subjective experience. The realm of "things in themselves" (*noumena*) describes external things existing independently of our experience, including any non-self minds.

We lack access to the internal world of plants and cannot reliably attribute intrinsic ends. And yet we must act as though such ends exist. The power, form, and function of life can only be understood in terms of a final cause, despite our inability to establish such a cause through observation or reason (Chaouli 2017). Kant still names the Aristotelian ends of nutrition, growth, and reproduction.

For Kant, final causes do not exist in competition with efficient causes. Rather, they are a different way of thinking about the same thing. In

---

[2] Mix 2018, pp. 163–164; see Kant's *Critique Judgment*, part 2.

Kantian language, a living thing is "both cause and effect of itself." This idea echoes Aristotle's confluence of efficient and final causes and will return in the autocatalysis and autopoiesis of twentieth-century biologists. More importantly, his perspective opened the door for the recursive process of natural selection as environments and genes shape one another through intermediary phenotypes and populations.

Organism Versus Mechanism

Following Descartes, the mechanical philosophers distinguished life and agency from simple matter. In Medieval thought, God acted in nature, but in Enlightenment thought, God acted on nature. Divine final causes competed with efficient mechanical causes. Biologists, philosophers, and theologians came to see biological activity requiring *more* divine intervention than simple physical activity.

John Stuart Mill (1885, 8) pointed out the confusion. We can use "nature" in two ways. We might include all things and all powers. Or we might include only those things that happen without will and intent. Consider the distinction between natural and artificial selection. If we make them mutually exclusive, we deny the natural-ness of human action. Mill focused on human power, but views on God's power changed as well. The modern definition of miracle—as unnatural intervention by God's will—also solidified in the nineteenth century.[3] Intelligent design arguments (e.g., Paley's *Natural Theology*, 1809) relied on this distinction between natural and artificial causation. If nature excludes power, form, and function a priori, then it excludes biology. Alternatively, if nature includes biology, it must include power, forms, and function.[4]

---

[3] *Oxford Dictionary of the Christian Church* (2005), s.v. "miracle."

[4] Parallel arguments can be found in astronomy. Johannes Kepler argued that the geometric regularity of the cosmos revealed divine nature and intelligence (*Mysterium Cosmographicum*, 1597). He would later argue that the circular character of craters on the Moon must be the work of intelligent natives. Isaac Newton believed that the distribution of light and dark matter in the solar system required a "voluntary Agent" (*Original letter from Isaac Newton to Richard Bentley, dated 10 December 1692*). Two centuries later astronomer Percival Lowell (1855–1916) argued for the existence of life on Mars based on linear formations or "canals" reported by Schiaparelli (*Mars as the Abode of Life*, 1908). All three considered "nature" insufficient to explain such regular features and concluded that intelligence must be involved. These arguments reveal a long tradition of distinguishing between nature and intelligence within the natural sciences.

Bacon's epistemological dualism and Descartes' ontological dualism broke biology. Bodies became natural, while power, form, and function became unnatural and unknowable. Biologists faced a choice. They could accept an unnatural biology or redefine nature. Biological actors diversified in response to the choices of individual thinkers. The insights of Kant, Darwin, and the Modern Synthesis would eventually produce natural biological agents (i.e., genes and populations) through a coevolution of "nature" and "teleology."

THE EFFECTS OF EVOLUTION

A second factor exacerbated the failure of Aristotelian biology: progress. Developmental theories about the universe have a long history (Ruse 2010, pp. 11–53). Plato explained vegetable life with vegetable souls that unified and motivated bodies (Lorenz 2006; Mix 2018, pp. 29–41). As part of a cosmic soul, they participated in the cosmic movement from becoming to being (Buchheim 2006). This made the entire cosmos one organism. Medieval scholars came to interpret this as an eternal (atemporal) positive movement from base matter to Divine Intelligence. They called it the *Scala Naturae* (literally, the ladder of nature) or great chain of being (Lovejoy 1971).

In the eighteenth century, the static chain mutated into a theory of historical progress (Lovejoy 1971, chapter 9; Ruse 1996, pp. 42–49; Sloan 2017). This required two important innovations. The first was the opposition of will and nature discussed above. Many thinkers, including scientists, saw biological power, form, and function as evidence of an external force acting on passive, mechanical nature. The effects of this action, particularly over long periods, could reveal divine intentions. If God's will could not be known directly, it could be discovered through the effects of God's actions. The second innovation was change through time. Medieval biologists thought of a static array. Earlier thinkers would have found biological progress unthinkable due to the eternal fixity of species in the mind of God. By rejecting transcendent ends, Bacon and Descartes opened the way for historical change.

"Evolution" and Biology

At first, "evolution" referred to organismal development. The meaning changed in the mid-eighteenth century. Charles Bonnet (1720–1793) used it for his theory of *preformationism*. Bonnet argued that God pre-made all living things and implanted seeds within seeds within the first organisms. Successive generations revealed, literally unrolled, the entire plan. Bonnet and Jean-Baptiste Robinet (1735–1820) linked the power that directed individual development to a greater power directing long-term changes in species. Biologists began to talk about the forces driving evolution. The ladder of nature became an escalator. Species became objects in time, with a beginning and an end. Form and function became flexible and, potentially, natural.

Progressive theories dominated (Ruse 1996; Sloan 2017). Biologists retained the old hierarchy of value so that evolution proceeded from simple to complex, passive to active, and base to dignified. Expert opinion favored a fixed trajectory for life, running from "lower" to "higher" organisms. Jean-Baptiste Lamarck (1914, pp. 56–61) referred to it as the "order of nature" and path to perfection. John Locke (1824, pp. 483–4) and George-Louis Leclerc, Comte de Buffon (1797, pp. 255–272) defended a single continuum with animals arising from plants. Robert Chambers (1844, p. 191) and Lamarck (1914, p. 51) proposed distinct plant and animal trajectories. Theories ranged from agnostic to theist, but all agreed on empirically observed progress.[5]

Darwinian Biology

Charles Darwin (1809–1882) suggested that biology could be neither transcendent nor progressive. His *Origin of Species* (1876) documented rampant extinction, divergence of character, and fitness to local conditions, all of which argued against Lamarck's path to perfection. Natural selection was less ambitious than earlier forms of evolution. It produced local adaptation, not global improvement. It grounded form and function in historical environments and populations. Darwin traced the development of form and function, but never found a satisfactory account of the biological actors that empower, inform, and direct the process.

Darwin rejected divine agency and will in evolution. Early in his career, he followed the lead of Francis Bacon and Erasmus Darwin. Transcendent order may be necessary for the general operation of the physical world; it is not uniquely necessary for biology (Mix 2018, pp. 194, 200–201; Ruse

---

[5] Agnostic, Herbert Spencer's "inherent tendencies"; deist, Lamarck's "blind force"; and theist, Asa Gray's "theistic evolution."

1996, p. 166). Later in his life, Charles Darwin became more clearly agnostic. After discovering natural selection, he never saw God as a useful part of biological explanation.

Natural selection provided a way of discussing change through time that did not rely on a mental teleology, associated at the time with Aristotelian final causes.[6] In its place was a causal nexus, similar to Kant's interpretation. Darwin explicitly addressed arguments against illicit teleology and vitalism, starting with the third edition of *Origin*. Natural selection was a fitness focused—and therefore ends focused—way of thinking about a long chain of efficient causes.[7] It has the power to do work—like a vital force—without being unnatural.[8] He also tackled the Medieval argument that plants cannot have minds (specifically will) and thus cannot have the finality of natural selection.[9]

Darwin critiqued both linearity and goodness in evolution. His branching tree replaced the fixed trajectory of earlier evolutionists. Life diversifies. Species develop in different ways and may even become less complex (Ruse 1996, pp. 147–150). This made change possible, if neither

---

[6] Darwin (1876, p. 383): "Nothing can be more hopeless than to attempt to explain this similarity of pattern in members of the same class, by utility or by the doctrine of final causes. The hopelessness of the attempt has been expressly admitted by Owen in his most interesting work on the 'Nature of Limbs.' On the ordinary view of the independent creation of each being, we can only say that so it is;—that it has pleased the Creator to construct all the animals and plants in each great class on a uniform plan; but this is not a scientific explanation. [¶] The explanation is to a large extent simple on the theory of the selection of successive slight modifications,—each modification being profitable in some way to the modified form, but often affecting by correlation other parts of the organisation."

[7] Ibid., p. 63: "So again it is difficult to avoid personifying the word Nature; but I mean by Nature, only the aggregate action and product of many natural laws, and by laws the sequence of events as ascertained by us."

[8] Ibid. p. 514: "It is no valid objection that science as yet throws no light on the far higher problem of the essence or origin of life. Who can explain what is the essence of the attraction of gravity? No one now objects to following out the results consequent on this unknown element of attraction; notwithstanding that Leibnitz formerly accused Newton of introducing 'occult qualities and miracles into philosophy.'"

[9] Ibid. p. 63: "Others have objected that the term selection implies conscious choice in the animals which become modified; and it has even been urged that, as plants have no volition, natural selection is not applicable to them! In the literal sense of the word, no doubt, natural selection is a false term; but who ever objected to chemists speaking of the elective affinities of the various elements?—and yet an acid cannot strictly be said to elect the base with which it in preference combines."

inevitable nor uniform. Darwin also highlighted biological nastiness, most famously in parasitic wasps (Darwin 2018).

Darwin's overall views on progress remain controversial (see Ruse 1996, pp. 136–177). Throughout his life, he displayed keen interest in the difference between "higher" and "lower" forms of life, a continuum that seemed clear to him. His treatment of human evolution in *Descent of Man* and elsewhere reveals an attachment to progress both in the rise of humans and in the development of human societies. Elsewhere, he spoke strongly against biological progress. More importantly, he provided a conceptual framework—natural selection—in which biological agents need not be driven toward progress in any absolute sense. They were free to adapt to local environments and adopt strategies that seemed inconsistent with providence and human ethics.

*   *   *

Biologists from the seventeenth century onward—whether labeled natural philosophers, physicians, physiologists, or biologists—cared deeply about power, form, and function and how they fit with nature. Accusations of "vitalism" and "teleology" reveal concerns that the theories of others failed as natural science. They also reflect competition over who would define natural science and the bounds of "nature" as a concept. This fundamental category shifted along with the field of biology and the range of biological agent concepts.

The mechanical philosophy narrowed nature to matter in motion, but it remained unclear whether life fit within such a world. Human life remained safely abstracted, moved by immaterial rational souls, but vegetable and animal souls no longer had meaning. They relied too heavily on extrinsic form and function, formal and final causes now thought to transcend nature and human knowing. New concepts would have to work within the confines of a physical world, though the exact limits of that world were still up for debate.

Evolutionary perspectives created a new view of life, in which biological agents moved up the ladder of nature. "Higher" life forms either came to light or emerged from "lower" life forms over time. Such mobility opened up new ways of thinking about power, form, and function, particularly the fixity and potential ends of species. Early proponents of evolution saw an overall trajectory of life on Earth, progress toward perfection. Darwin's theory of evolution by natural selection presented an alternative—a natural method for adaptation to local environments based on a history of competition for resources. He located biological power in a complex interaction of populations and environments. He also explained how a form and function might arise through a recursive process of inheritance, variation, and selection. Individual agents, however, remained an open question. What are the units of inheritance and selection? And how do power, form, and function reside in them?

The next chapter looks at how the vegetable souls of "Aristotelian" biology speciated into diverse biological actors, notably preformed embryos, vital particles, vital forms, and vital forces. All four would eventually contribute to modern biology, through recombination and symbiosis, resulting in the genes and populations familiar to twenty-first-century biologists. These new biological actors are natural, context-specific, and blind, following the developments of natural science and natural selection. Their "agency" depends on interactions with one another in a particular environment.

References

Bacon, Francis. *The Advancement of Learning*, edited by Michael Kiernan. *The Oxford Francis Bacon*, volume 4. Oxford: Oxford University Press, 2000.

Buchheim, Thomas. "Plato's *phaulon skemma*: On the Multifariousness of the Human Soul." In *Common to Body and Soul*, edited by R.A.H. King, 103–120. New York: Walter de Gruyter, 2006.

Buffon, Georges Louis Leclerc. "Barr's Buffon." In *Buffon's Natural History, Containing a Theory of the Earth, a General History of Man, of the Brute Creation, and of Vegetables, Minerals, Etc. From the French*, 10 vol., translated by J. S. Barr. London: H.D. Symonds, 1797.

Chambers, Robert. *Vestiges of the Natural History of Creation*. London: John Churchill, 1844.

Chaouli, Michel. *Thinking with Kant's "Critique of Judgment."* Cambridge, MA: Harvard University Press, 2017.

Darwin, Charles. *On the Origin of Species by Means of Natural Selection, or the Preservation of Favoured Races in the Struggle for Life* 6th ed. London: John Murray, 1876.

Darwin, Charles. 2018. Letter no. 2814. Darwin Correspondence Project. http://www.darwinproject.ac.uk/DCP-LETT-2814. Accessed 7 February 2018.

Klein, Jürgen. "Francis Bacon." In *Stanford Encyclopedia of Philosophy*, Winter 2016 ed. Stanford University, 1997–. https://plato.stanford.edu/archives/win2016/entries/francis-bacon/.

Lamarck, Jean-Baptiste. *Zoological Philosophy: An Exposition with Regard to the Natural History of Animals.* Translated by Hugh Elliot. London: Macmillan, 1914.

Locke, John. *The Works of John Locke in Nine Volumes,* 12th ed. London: Rivington, 1824.

Lokhorst, Gert-Jan. "Descartes and the Pineal Gland." In *Stanford Encyclopedia of Philosophy,* Winter 2017 ed. Stanford University, 1997–. https://plato.stanford.edu/archives/win2017/entries/pineal-gland/.

Lorenz, Hendrik. *The Brute Within: Appetitive Desire in Plato and Aristotle.* New York: Oxford University Press, 2006.

Lovejoy, Arthur. *The Great Chain of Being.* Cambridge, MA: Harvard University Press, 1971.

Mill, John Stuart. *Nature, the Utility of Religion, and Theism,* 3rd ed. London: Longman, Green, 1885.

Mix, Lucas J. *Life Concepts from Aristotle to Darwin: On Vegetable Souls.* New York: Palgrave, 2018.

Osler, Margaret J. *Divine Will and the Mechanical Philosophy.* New York: Cambridge University Press, 2004.

Osler, Margaret J. *Reconfiguring the World: Nature, God, and Human Understanding from the Middle Ages to Early Modern Europe.* Baltimore: Johns Hopkins University Press, 2010.

Paley, William. *Natural Theology,* 12th ed. London: J. Faulder, 1809. http://darwin-online.org.uk/.

Rees, Graham. "Atomism and 'Subtlety' in Francis Bacon's Philosophy." *Annals of Science* 37, no. 5 (1980): 549-571.

Ruse, Michael. *Monad to Man: The Concept of Progress in Evolutionary Biology.* Cambridge, MA: Harvard University Press, 1996.

Ruse, Michael. *Science and Spirituality: Making Room for Faith in the Age of Science.* New York: Cambridge University Press, 2010.

Sloan, Phillip. "The Concept of Evolution to 1872." In *Stanford Encyclopedia of Philosophy,* Spring 2017 ed. Stanford University, 1997–. https://plato.stanford.edu/archives/spr2017/entries/evolution-to-1872/.

CHAPTER 5

Who "Acts" in Biology? Biological Agents from Souls to Genes

Abstract Living things display unique power, form, and function. Ancient and Medieval biologists explained this with appeals to vegetable souls and final causes. By the seventeenth century, these explanations were viewed as unacceptably mental and supernatural. The vegetable soul concept mutated and diversified over the next three centuries into a menagerie of biological "agents," including preformed embryos, vital particles (e.g., gemmules, plastidules, biophors, and ids), vital forms (e.g., *moule intérieur*), and vital forces (e.g., mesmerism and galvanism). Reciprocal accusations of "vitalism" and "teleology" reflect competition over who would define nature and natural science. Biologists considered and rejected three kinds of teleology—transcendent, progressive, and prospective—leading to selection among the agents. Internal drivers (orthogenesis) and vital forces flourished briefly, but ultimately died out. Population genetics arose through a recombination of the two, leaving genes as the only extant representative. The phylogeny of biological actors reveals the conceptual history of biology and why some of the current philosophical issues appear intractable. Modern concepts, particularly genes, organisms, and species, arose through a selection process. Knowing that process can tell us how they worked historically and whether they might work in the future.

Keywords Gene • Orthogenesis • Nature • Teleology • Vegetable soul • Vitalism

© The Author(s), under exclusive license to Springer Nature Switzerland AG 2022
L. J. Mix, *The End of Final Causes in Biology*,
https://doi.org/10.1007/978-3-031-14017-4_5

Vegetable souls never really died; they evolved. This chapter covers the range of biological actors or "agent" concepts that inherited power, form, and function from vegetable souls in the early modern period as biologists attempted to reconcile the dynamic consistency of living things with new understandings of nature. From vital particles (such as gemmules and plastidules) to vital forces (such as mesmerism and animal magnetism) to orthogenesis and the *moule intérieur*, these concepts were proposed as sincere attempts to make biology more scientific. And, though most are now extinct (as biological concepts), a few passed on language and conceptual framing to genes. That history reveals why modern gene concepts work the way they do. It shows how power, form, and function still shape biology.

Aristotle's concept of the vegetable soul might have been compatible with natural science. He located power, form, and function in the process of nutrition (Chap. 2). In action, and in fulfillment, this process was inherently mortal and material. Over the centuries, however, souls had become something else entirely. Associated with subjectivity and spirituality, the human soul came to be identified with the mind, an immaterial and transcendent agent (Chap. 3). Form and function were associated with such minds. Nature, meanwhile, came to be viewed as nothing more than matter in motion: mindless, passive, and mechanical (Chap. 4). Vegetable and animal souls could not be fit into either category.

This left biologists in a bind. The dynamic consistency of living things requires power, form, and function beyond what appears in simple matter. Plants and non-human animals clearly lack the kind of minds present in humans. How, then, do they have the "agency" necessary for nutrition, growth, and reproduction?

Medieval scholars had solved this problem by allowing all souls to participate in God's agency. Form and function resided in the mind of God, who informed and directed the cosmos through a continuum of increasing power and dignity, the ladder of nature. Enlightenment thinkers, by forcing a rigid separation between natural material bodies and transcendent ideal minds, eliminated this connection.

Biologists responded to this body-mind dualism in one of two ways, seeking to keep plant and animal life within the natural world. They could naturalize teleology, explaining how power, form, and function were not mental at all. They would need a theory of biological ends that avoided the formal and final causes of Medieval "Aristotelianism" and expressed ends as matter in motion. Alternatively, they could finalize nature, widening the empirical universe beyond the mechanical limits set by Descartes and others. This would allow "agents" and "ends" back in. Biology as a field did a little of each.

This chapter follows the proliferation of quasi-natural proto-agent concepts used by early modern biologists. Most viewed their work as a form of natural science. In retrospect, they appear to be "vitalist"—improperly teleological—because they transgress the currently understood bounds of nature. Those bounds, however, were in flux. Biologists were attempting to discover the fundamental patterns of biology, but they were also competing with one another over whose model of nature would win out.

Biology as a field embraced a range of teleologies. Vegetable souls continued to occupy a conceptual niche through the end of the nineteenth century. The restrictions of modern science continued to adapt. Internal driver concepts (e.g., orthogenesis) and vital force concepts (e.g., *élan vital*) flourished briefly, but ultimately died out. The power they attempted to address was shifted to interactions between population and environment. Their prospective and progressive aspects were rejected (see Chap. 6). Vital particle concepts (e.g., gemmules) and vital form concepts (e.g., the *moule interieur*) passed on their teleology to a new biological agent, the gene. Localized to nucleic acids, the gene as molecule could be a physical particle and a mechanical locus for information. And yet, it was not the single molecule that acted (a gene token). Instead, a collection of molecules with the same sequence, a new immaterial particle eventually took on aspects of power, form, and function (a gene type or "strategic gene").

Biological Actor Concepts Proliferate

Vegetable souls evolved into at least four populations: preformed embryos, vital particles, vital forms, and vital forces. Each concept provided a locus for power, form, or function as biologists attempted to establish a new understanding of biology that aligned with natural science. Many early modern biologists advocated for multiple concepts. Buffon, for example, is commonly associated with the *moule intérieur* (a vital form) but he also appealed to vital particles (for biological power).

Preformed Embryos

Two competing naturalisms arose within biology (Maienschein 2017). Epistemological materialists started with empirical observation of organisms arising and changing. They argued that matter could carry form and function and use it to inform other matter, a process they called *epigenesis*.[1] Stored form (or information) could be used to make new individuals (generation), change individual form and function through time (development), and restore form and function when pieces were lost (regeneration). This fit well with both Aristotle and Medieval Aristotelianism but offended against the mechanism of Descartes.

Ontological materialists objected to epigenesis, finding it dangerously vitalist and teleological. They argued that material mechanical bodies cannot contain form and function, much less impose them on other matter. The disagreement paralleled Mill's two uses of nature. Epigeneticists observed biological powers and looked for "natural" explanations. Their opponents defined "nature" and excluded biological powers.

Pioneers of microscopy suggested an alternative to epigenesis: *preformationism*. Antoni von Leeuwenhoek (1632–1723) documented organization and movement at the microscopic level. Marcelo Malphigi (1628–1694) and Jan Swammerdam (1637–1680) found morphological complexity in embryos. Nicolaas Hartsoeker (1656–1725) and Johan Ham (c.1651–1723) observed sperm for the first time. All of them argued that large organisms grow from microscopic versions (in humans, *homunculi*).

Preformationists diverged over the means of transmission. *Spermists* argued that these preformed embryos passed from father to child by way of sperm (e.g., Leeuwenhoek, Hartsoeker, Ham, Leibniz, Boerhaave). *Ovists* argued that embryos passed from mother to child by way of eggs (e.g., Malphigi, Haller, Bonnet). Preformationists were the first proponents of "evolution" as the gradual appearance of new kinds as the scroll of life unrolled (Latin: *evolvere*). They dominated the iatromechanist school, proponents of mechanical thinking in medicine.

---

[1] Epigenesis as a theory of forms refers to forms changing through time and new forms arising, in contrast to preformationism. This was a primary concern in eighteenth-century biology and decided in favor of the epigenesists. Epigenetics as an exploration of gene expression and development, including the possibility of non-DNA loci of inheritance, was a distinct, if related, issue of concern in the twentieth century.

Preformationism also impacted philosophers of biology.[2] Nicolas Malebranche (1683–1715) denied internal agency in organisms (Malebranche 1980, pp. 492–496). He explicitly used preformation to naturalize form and function (Detlefsen 2003). God is the only true agent; no true reproduction occurs, only the successive appearance of embryos. Malebranche accepted the vegetable soul, but only as a way of speaking about "a corporeal motivator and complex behavior analogous to the spring and mechanism of a clock" (Mix 2018, p. 167). Wilhelm Gottfried Leibniz (1646–1716), on the other hand, believed in a miraculous correspondence between the material embryo/organism and immaterial monads (Mix 2018, pp. 162–163). He placed vegetable souls in the second category.

If God packed embryos, one within another, at the beginning of time, biology could proceed mechanically from then on. Divine power created all form and function in creation. In each generation, a new set of embryos are revealed, each one carrying the embryos of future generations. While invocation of God was not strictly necessary, preformationism simply deferred questions of power, form, and function to an earlier time. They did not resolve how biological forms could arise or persist as matter in motion.

Pioneering physiologist Albrecht von Haller (1708–1777) explicitly argued that divine action was necessary to explain biological organization. This led him to reject the epigenesis of Buffon in favor of preformationism. Organism requires "a directive intelligence, in order to place the organic particles appropriately" (Zammito 2017, p. 132). Contemporaries Pierre Louis Moreau de Maupertuis (1698–1759) and Caspar Friedrich Wolff (1733–1794) critiqued preformationism for precisely this reason. One a philosopher and the other a physiologist, both claimed that science sought natural causes of phenomena. Both considered God necessary as the ultimate cause of nature itself; they simply argued that biological teleology could be explained within nature through appeals to observation and reason (p. 154).

Preformationism waned in the eighteenth century and was largely extinct by 1900. As microscopy improved, biologists could see that

---

[2] Current disciplinary bounaries did not form until the nineteenth and twentieth centuries. Thinkers now known as philosophers and theologians—such as Malebranche and Kant—did experiments and wrote as natural scientists. Thinkers now known as scientists—such as Newton and Boyle—wrote on philosophy and theology.

embryos change shape as they develop. They also documented cases of regeneration and embryo splitting. Meanwhile scientific opinion shifted from theories of infinitely divisible matter to theories of atomism, making infinitesimal embryos inconceivable. Epigeneticists and epistemological materialists won the day and returned to Aristotelian concepts of generation and development. They painted preformationists as vitalist and anti-mechanist for their appeals to God, everlasting embryos, and invisible causes.

Proponents of epigenesis, in turn, had to explain how power, form, and function could be explained without an ultimate cause, divine or otherwise. How could they locate teleology in wholly natural biological agents?

Vital Particles

A second species of biological actor concept arose as vital particles. Aristotle (1936) had argued that a dynamic cause was needed to explain the dynamic behavior of living things: self-movement and spontaneous motion. Neither a materialist nor an atomist, he attributed this self-movement to souls. Lucretius (1922, books III and IV) wanted something closer to an atom or element. He invoked active, unpredictable *soul seeds* as components of corporeal souls. Some biologists, looking for alternatives to Aristotle, turned to Lucretius.

Many eighteenth-century biologists located power, information, and/or function in particles—or in fluids composed of particles. Buffon (1707–1788) excluded divine causation, final causes, and souls but argued for *organic particles* with vital powers (Buffon 1797, pp. 301–309; Sloan 2017). Expressly anti-mechanist, he nonetheless saw organic particles as more natural than souls. Erasmus Darwin (1731–1802) spoke of *subtle spirits*, matter with intrinsic ("primary") motion that animated life (Darwin 1794, p. 5–6). Jean-Baptiste Lamarck (1744–1829) wrote of *subtle fluids* comparable to electricity that empowered nutrition, growth, and reproduction (Lamarck 1914, pp. 187–188). Lamarck took pains to note that these fluids are physical but have life-specific activities. They followed Descartes, who spoke of rarified material *spirits* (Osler 1994, p. 220), but proposed a broader naturalism with active particles.

Many early proponents of "evolution" sought to supplement (or supplant) natural selection with innate drivers that could direct or determine the path of change: *orthogenesis* (Bowler 2017). Modest versions referred only to chemical or organizational constraints on variation. More dramatic versions returned to fixed pathways for species development. Orthogenesists still claimed to reject "teleology" when they denied divine intervention and progress. Some even used "negative" trends as proof of their ortho-doxy (e.g., the *dysteleology* of Haeckel).

Orthogenetic particles had agent-like power to "carry" hereditary information and "direct" chemical reactions. Charles Darwin proposed self-dividing *gemmules* for inheritance (Darwin 1875, p. 372; Winther 2000). Ernst Haeckel (1834–1919) attributed memory and agency to *plastidules* (Haeckel 1876; Uschmann 1979). August Weismann (1893) gave vital powers to *biophors* and *ids*. Leonard Troland (1914) allowed self-reproduction and growth in *autocatalytic enzymes*.

Proponents of vital particles argued that they were necessary to natural biology. Material and, in theory, observable particles would be amenable to reductionism and mechanical explanations, if not ontological material-ism. Souls could be reframed as dynamic assemblies of dynamic atoms. Around 1900, biologists began to identify these particles with observable physical units (Troland 1917; Muller 1966), retaining the language of order and agency. They shifted biological agency up a level in the material hierarchy, from fundamental particles to molecules (themselves composed of generic atoms). Genes and proteins took on the language of power, form, and function as conceptual descendants of these molecular vital particles.

Vital Forms

Vital particles explained the power aspect of living things but were less appealing in explanations of form and function. Many who wanted them to be fundamental particles could not easily explain how they provided different species with different traits. Drawing on the pneumatic theories of the Stoics, some biologists sought to replace the formal cause aspects of vegetable souls with a physical mold or template. The most famous of these may be Buffon's internal mold (*moule intérieur*). Lamarck used the term *orgasm* with similar meaning.

Like earlier corporeal theories of the vegetable soul, these vital forms could be used to organize vital bodies out of components: generic parti-cles, vital particles, or some combination of the two. Configurations of particles might then have emergent properties, making them consistent with physics without being limited by physical description. Many viewed this as a minor but acceptable adjustment to the mechanical philosophy.[3]

In the twentieth century, genes and enzymes took on properties of vital forms as well as those of vital particles. Like particles, they were mathematically unitary and fundamental. Like vital forms, they were physically composite and capable of encoding traits. Genes became nucleic acids that carry information, causal loci above passive physics but below active biology (Keller 2002, pp. 123–132). Enzymes became polypeptide sequences that direct chemical reactions. Proteins, more generally, became polypeptides with biological function.

Early twentieth-century biologists differentiated them as agents but distinguished them from earlier agent concepts by rejecting prospect or intention. Troland (1914), for example, referred to enzymes as regulators, creators, assistants, mediators, duty-bound synthesizers, controllers, aids, directors, facilitators, builders, and guides of function—all within a single article. He identified them with earlier biological actors, notably Weismann's biophors and determinants (p. 118), but considered them to be "a strictly physico-chemical mechanism" for explaining life (p. 131). He drew the line at "purposeful effort on the part of an individual," which he identified with vitalism, teleology, and orthogenesis (p. 107).

The critical question was not *if* particle/forms can be biological actors, but *how*. In the early twenty-first century, biologists still speak of genes having power, carrying information, and imbuing function in ways that abiological molecules do not.

Vital Forces

Many early biologists were inspired by the systematic approach to physics that unified abiological causes with appeals to natural forces—most notably electricity and magnetism. Such a force for biology could power the whole system. While it might not provide form and function at the level of organs, it could—if real—explain biological dynamism and an overall direction for organisms and evolution.

---

[3] The vital forms inherited from Stoicism relied on a side-by-side materialism of nature breath, with a rarified body permeating a grosser, tangible body (Mix 2018, pp. 70–71). While this may seem unnatural by modern standards, it has an interesting parallel in the contemporary theory of luminous ether, the medium through which light was believed to travel.

Some biologists appealed to electricity and magnetism or similar natural forces. Others appealed to a transcendent *Spirit*, revealing the influence of German Transcendentalism on biology (Ruse 1996, p. 181; Miller 2002). Both groups frequently contrasted their ideas with the "vitalism" and "teleology" of vital particles and forms.

The history of vital forces deserves special attention in the broader history of science. The earliest proponents of the mechanical philosophy believed efficient causes required physical contact. There could be no action at a distance. Astronomers in the seventeenth century invoked invisible forces for precisely this reason—to expand the bounds of "nature." Kepler and Newton used magnetism as an analogy for the attraction of gravity.[4] Newton also spoke of inertia as an innate force (*vis insita*). Seventeenth-century physicists and chemists began to differentiate electricity and magnetism.[5] It is unsurprising to find biologists in the eighteenth century attempting similar explanations and considering them fully natural.

In the century before Darwin, many biologists appealed to life-specific forces and imagined a vital naturalism, governed by universal laws but including a driving force particular to biology. Wolff (1733–1794) looked for an essential force (*vis essentialis*). Franz Mesmer (1734–1815) suggested *animal magnetism* (or *mesmerism*). Luigi Galvani (1737–1798) and Lorenz Oken (1779–1851) used *animal electricity* (or *galvanism*). Johann Friedrich Blumenbach (1752–1840) coined the term *Bildungstreib*, German for driving force, entirely distinct from general physics "which appears to be one of the original causes of all propagation [Generation], nutrition, and regeneration [Reproduction]" (Zammito 2017, p. 212, bracketed terms from Zammito). This idea became popular among nineteenth-century biologists as a guiding hand for both individual development and progressive evolution ("development") of species.

After Charles Darwin, some maintained these forces alongside natural selection. Herbert Spencer (1820–1903) thought evolution produced observable progress from homogeneity to heterogeneity (Spencer 1857; Ruse 1996, pp. 181–191). He attributed this to the interaction of

---

[4] Johannes Kepler, *Astronomia Nova* (New Astronomy) 1609; Isaac Newton, *De Motu Corporum in Gyrum* (*On the Motion of Bodies in Orbit*) 1684 and *Philosophiæ Naturalis Principia Mathematica* (*Mathematical Principles of Natural Philosophy*) 1687.

[5] William Gilbert, *De Magnete, Magneticisque Corporibus, et de Magno Magnete Tellure* (*On the Magnet and Magnetic Bodies, and on That Great Magnet the Earth*) 1600.

individuals with their environment (Spencer 1864, p. 430) but explicitly denied vitalism, which he defined as innate teleology imposed by a supernatural actor (pp. 403–404). Alternatively, Alfred Russel Wallace (1823–1913) argued for a transcendent *Spirit* that intervened in evolution to bring about nutrition, sensation, and reason (Wallace 1889, pp. 472–474). He said that life requires both "directive agency" and "organizing power" (Wallace 1910, p. 333).[6]

Henri Bergson (1859–1951) defended the creativity and contingency of evolution against determinism (Bergson 1998, p. 96). He opposed both strict mechanism and end-oriented teleology or *finalism*. Instead, he embraced a pervasive driver, *élan vital* (Lawlor and Moulard-Leonard 2016). Pierre Teilhard de Chardin (1881–1955) spoke of progress toward complexity, consciousness, and unity (Teilhard 1959). All three required a biological actor that pervaded nature and drove biology forward.

Comparison with physics may again be useful, as in the case of Newton's Law of Universal Gravitation and the Second Law of Thermodynamics.[7] Universal directional forces sound inherently unnatural in the context of modern biology. They have, however, been successful in physics (e.g., increasing entropy) and could not be ruled out a priori. They had to be excluded by observation as we will see in Chap. 7.

While intrinsic vital drivers were labeled orthogenesis, extrinsic or universal vital drivers attracted the label "Lamarckism" (Bowler 2017). Lamarck favored vital particles and forms but spoke compellingly about individuals interacting with their environment. These interactions led later biologists to refer to theories of evolution driven by environmental forces as "Lamarckian" when they contrasted vital forces and natural selection. Few would have been embraced by Lamarck, himself. Modest versions of Lamarckism only emphasized environmental and epigenetic processes.[8] More dramatic versions returned to progress and Divine intervention.

---

[6] For comparison, Descartes, Newton, and Galvani spoke of signals conducted from senses to brain and brain to muscles via "subtle spirits" akin to Stoic breaths, but also akin to modern electrical signals.

[7] Newton introduced his Law of Universal Gravitation in the *Principia*. The Second Law of Thermodynamics developed gradually but was clearly stated by Rudolf Clausius and William Thomson, Lord Kelvin around 1850.

[8] Lamarckism has been associated with all three kinds of "epigenesis," the historical refutation of preformationism and location of biological power within nature, gene expression and development, and non-Darwinian evolution.

Interaction between individuals and environmental processes proved central to Darwin's theory and the Modern Synthesis. The "genes" that emerged as the dominant biological actor did work in the context of a genetic population changing within a definite environment over a definite period of time. Vital forces fell out of fashion in the early twentieth century, but the role of the environment took center stage. Biologists began to speak of selection as a force and asked about competing forces. Biological power, information, and function could not be located in genes themselves (orthogenesis) nor in environmental forces alone (Lamarckism) but only in interactions between the two. The leaders of the Modern Synthesis maintained both particles and forces as biological actors but scrubbed them of prospect. They insisted that evolution was random, contingent, and blind.

*   *   *

Vegetable souls never truly disappeared. Like the dinosaurs, they evolved into something new. Their evolution reveals important continuities and discontinuities in biology. The category of "life" still describes a distinct set of physical processes and structures characterized by unique power, form, and function. Those descriptions, whether objective or merely instrumental, rely on invocations of genes as discrete, immaterial actors with ends of their own. In this way, modern discussions of biological individuality, species, and functions reflect long-standing questions about the relationship between life and non-life. Biology has not escaped the problems described by Aristotle in *On the Soul*. If anything, modern research has made it once again critical to consider the dynamic consistency and strange naturalness of living things.

Vegetable souls, under that name, had a last gasp of popularity among the animist physicians of the early eighteenth century. Their most famous proponent was Georg Stahl (1659–1734). Later physiologists and biologists avoided the term, though Hans Driesch (1914, p. 80) clearly echoes the concept with "non-mechanical suprapersonal determining agents" or *entelechy*. Philosophers and theologians retained the term "soul" but spoke almost exclusively of human souls.

Careful attention to the evolution of terms reveals that most students of life from the seventeenth century onward—whether labeled natural philosophers, physicians, physiologists, or biologists—cared about power, form, and function and how they fit with nature. Accusations of "vitalism" and "teleology" reflect competition over who would define nature and natural science.

The next chapter turns to genes and how they were used to explain power, form, and function in the early twentieth century. They took on the mantle of primary biological agent (concept) by situating power, form, and function in embodied organisms. Incorporating aspects of vital particles and vital forms, they once again appealed to a recursive process of form informing matter.

References

Aristotle. *On the Soul, Parva Naturalia, On Breath.* Loeb Classical Library. Cambridge, MA: Harvard University Press, 1936.

Bergson, Henri. *Creative Evolution.* Mineola, NY: Dover, 1998.

Bowler, Peter J. "Alternatives to Darwinism in the Early Twentieth Century." In *The Darwinian Tradition in Context*, ed. Richard Delisle, 195–217. New York: Springer, 2017.

Buffon, Georges Louis Leclerc. *Buffon's Natural History*, vol. 2. Translated by J.S. Barr. London: H. D. Symonds, 1797.

Darwin, Charles. *Variation Under Domestication*, 2nd ed., vol. 2. London: John Murray, 1875.

Darwin, Erasmus. *Zoonomia*, 2nd ed., vol. 1. London: J. Johnson, 1794.

Detlefsen, Karen. "Supernaturalism, Occasionalism, and Preformation in Malebranche." *Perspectives on Science* 11, no. 4 (2003): 443–483.

Driesch, Hans. *The Problem of Individuality.* London: Macmillan, 1914.

Haeckel, Ernst. *Die perigenesis der Plastidule oder die Wellenzeugung der Lebenstheilchen.* Berlin: Georg Reimer, 1876.

Keller, Evelyn Fox. *Making Sense of Life.* Cambridge, MA: Harvard University Press, 2002.

Lamarck, Jean-Baptiste. *Zoological Philosophy.* Translated by Hugh Elliot. London: Macmillan, 1914.

Lawlor, Leonard, and Valentine Moulard-Leonard. "Henri Bergson." In *Stanford Encyclopedia of Philosophy*, Summer 2016 ed. Stanford University, 1997–. https://plato.stanford.edu/entries/bergson/.

Lucretius. *The Nature of Things.* Translated by William Leonard. Boston: E. P. Dutton, 1922.

Maienschein, Jane. "Epigenesis and Preformationism." In *Stanford Encyclopedia of Philosophy*, Spring 2017 ed. https://plato.stanford.edu/archives/spr2017/entries/epigenesis.

Malebranche, Nicolas. *The Search After Truth and Elucidations of the Search After Truth and Philosophical Commentary.* Translated by Thomas M. Lennon and Paul J. Olscamp. Columbus: Ohio State University Press, 1980.

Miller, Elaine P. *The Vegetative Soul: From Philosophy of Nature to Subjectivity in the Feminine.* Albany, NY: State University of New York Press, 2002.

Mix, Lucas J. *Life Concepts from Aristotle to Darwin: On Vegetable Souls.* New York: Palgrave, 2018.

Muller, Hermann Joseph. "The Gene Material as the Initiator and the Organizing Basis of Life." *American Naturalist* 100, no. 915 (1966): 493–517.

Osler, Margaret J. *Divine Will and the Mechanical Philosophy.* Cambridge: Cambridge University Press, 1994.

Ruse, Michael. *Monad to Man: The Concept of Progress in Evolutionary Biology.* Cambridge, MA: Harvard University Press, 1996.

Sloan, Phillip. "The Concept of Evolution to 1872." In *Stanford Encyclopedia of Philosophy*, Spring 2017 ed. Stanford University, 1997–. https://plato.stanford.edu/entries/evolution-to-1872/.

Spencer, Herbert. "Progress: Its Law and Cause." *Westminster Review* 67 (1857): 445–485.

Spencer, Herbert. *The Principles of Biology*, vol. 1. London: Herbert and Norgate, 1864.

Teilhard de Chardin, Pierre. *The Phenomenon of Man.* New York: Harper, 1959.

Troland, Leonard T. "The Chemical Origin and Regulation of Life." *The Monist* 24, no. 1 (1914): 92–133.

Troland, Leonard T. "Biological Enigmas and the Theory of Enzyme Action." *American Naturalist* 51, no. 606 (1917): 321–350.

Uschmann, Georg. "Haeckel's Biological Materialism." *History and Philosophy of the Life Sciences* 1, no. 1 (1979): 101–118.

Wallace, Alfred Russel. *Darwinism: An Exposition of the Theory of Natural Selection with Some of its Applications.* New York: Macmillan, 1889.

Wallace, Alfred Russel. *The World of Life: A Manifestation of Creative Power, Directive Mind and Ultimate Purpose.* London: Chapman and Hall, 1910.

Weismann, August. *The Germ-Plasm: A Theory of Heredity.* New York: Charles Scribner's Sons, 1893.

Winther, Rasmus G. "Darwin on Variation and Heredity." *Journal of the History of Biology* 33, no. 3 (2000): 425–455.

Zammito, John H. *The Gestation of German Biology: Philosophy and Physiology from Stahl to Schelling.* University of Chicago Press, 2017.

CHAPTER 6

Genes: The New Biological Agent

Abstract In the early twentieth century, genes arose as a new locus of power, form, and function. This new biological "agent" outcompeted other concepts to become a primary unit of biology, alongside organisms. The definition of genes, however, requires close attention. Single polynucleotides can be identified as discrete, material particles and units of form that can be passed from generation to generation. The real difference makers, however, require something more. A selfish or strategic gene that can be the "for whom" and "by whom" of biological function must be identified with a collective, a "type" of which individual polynucleotides are only "tokens." Recent gene concepts are both immaterial and inherently teleological. They allow biologists to retain teleology while avoiding the undesirable aspects of earlier transcendent, progressive, and prospective biological agent concepts.

Keywords Agency • Function • Gene • Teleology • Type/token

For Aristotle and many of his followers, vegetable souls situated life in material, corporeal reality. They were agents in the sense that they were meaningful difference makers. Humans, animals, and plants shared a common process of nutrition (in action and in fulfillment). They informed matter, changing what they eat into themselves. Consciousness and reason

L. J. Mix, *The End of Final Causes in Biology*,
https://doi.org/10.1007/978-3-031-14017-4_6

were encountered in that context. They were additional activities, even "higher" activities, but they were grounded in a material process.

Over the centuries, Western philosophers developed a "psychology" of subjective and spiritual life that blossomed in the Renaissance into a full-fledged dualism of body and mind. Enlightenment thinkers reified this divide into two distinct realms—one material, corporeal, and "natural" and one transcendent, mental, and "agential." Vegetable souls fell into the space between, along with the Aristotelian explanatory system of material, efficient, formal, and final causes.

Early modern biologists, hoping to develop biology as a natural science, sought to develop new biological "agents" that could take the place of the vegetable soul while adhering to the rules of empirical and rational reasoning. They largely rejected the strict mechanism of Descartes, seeking a vital naturalism that embraced teleology, if in a new way.

In the early twentieth century, genes arose as a new locus of power, form, and function. This new biological "agent" outcompeted other concepts to become a primary unit of biology, alongside organisms. The definition of genes, however, requires close attention. Single polynucleotides can be identified as discrete, material particles and units of form that can be passed from generation to generation. The real difference makers, however, require something more. A selfish or strategic gene that can be the "for whom" and "by whom" of biological function must be identified with a collective, a "type" of which individual polynucleotides are only "tokens." Recent gene concepts are both immaterial and inherently teleological. They allow biologists to retain teleology while avoiding the undesirable aspects of earlier transcendent, progressive, and prospective biological agent concepts.

BIOLOGY IN HISTORICAL PERSPECTIVE

Historians of biology rarely tell tales that start before Descartes and continue after Darwin. They have good reasons. Concept and context vary substantially with geography and time. The vegetable soul meant very different things to Aristotle, Lucretius, Aquinas, and Leibniz. It is unclear what, if anything, it would mean today. And so, stories of vegetable souls in natural philosophy became divorced from stories about the rise of modern biology. But Aristotle, Bacon, and Darwin all recognized "life" as a meaningful category. Each tried to understand the strange dynamism, complexity, and purposiveness of living things. By looking at descriptions of these traits in the seventeenth, eighteenth, and nineteenth centuries, we can track the evolution of concepts.

Just as dinosaur evolution helps us understand the traits of modern birds, so the evolution of soul concepts helps us understand the role of genes and populations in modern biology. The de-naturalization of souls in the Middle Ages and Enlightenment required a re-naturalization of biological "agents" afterward. The elimination of final and formal causes from nature required a re-finalization of matter and bodies.

Early modern scientists employed a peculiar menagerie of concepts, attempting to provide fundamental units for biology and loci for power, form, and function. These entities, which I have collectively labeled "biological agents" include *moule intérieur*, gemmules, plastidules, autocatalysts, and *élan vital*. No spectacular missing link connects souls to genes. Neither evolution nor history works that way. Instead, the menagerie gives us snapshots of specific times and places. We must look for common features that reveal the underlying tree of relationships. Rooting the tree in vegetable souls gives us tools for understanding both continuity and discontinuity in biological thinking.

The continuity comes from the desire for biological units and the three teleological traits attributed to them. Power describes the causal efficacy attributed to biological agents. From the time of Aristotle, this has been associated with the powers of nutrition (repurposing matter to be part of a biological body), growth (increasing the size or changing the shape of the body), and reproduction (making new biological bodies). Such power utilizes form, manifest, and hidden patterns that define the body and direct its operations. Holding these together, and making them teleological, is some end associated with both agent and body. An organism, literally and organized biological body, has parts working together toward a common end. The end of the whole determines the ends of the parts, their function.

The end of the whole might be considered a function, by analogy. It was viewed this way for over a thousand years, when the individual organism participated in a larger cosmic life. The mechanical philosophy (and nominalism) removed such transcendent ends, requiring a different view of organismal ends, still related to context but without the greater organism metaphor. Today, biologists speak of fitness or health, but they still invoke some idea that the whole organism has a proper end.

The end of the whole begs the question of what constitutes "the whole" or unit of teleology. For whom does the activity occur? The idea of a "selfish gene" or a "strategic gene" suggests that the gene is the primary beneficiary of selection and, thus, the appropriate locus of teleology. The history of biological agent concepts is a history of such loci.

I label vegetable souls, genes, and everything in between as biological "agents" to emphasize continuity. I do not intend to sneak in any ontological commitments about supernatural agency. Rather, I wish to highlight the way that discussions of power, form, and function predate modern distinctions between body and mind, between natural and unnatural. Indeed, they have been critical to creating those distinctions.

GENES

In the late eighteenth century, Gregor Mendel (1822–1844) proposed a mathematical treatment of inheritance, including discrete, independent units of information. In 1905 Wilhelm Ludvig Johannsen (1857–1927) coined the term "gene." Over the next 50 years, biologists gradually identified the genetic material—the material cause explanation of genes—as nucleic acids. The population of gene concepts, however, remained diverse, as biologists explored different bioteleologies and considered whether genes could be the repository for power, form, and function.

Genes as Form

Early genes were simple units of information, abstract packets of form that could be passed from one generation to the next. No concrete medium was known for storing them. Nor was there a mechanism for recording or expressing them. Text is just ink on a page (or pixels on a screen), but a word carries meaning from one interpreter to another. Genes were units of meaning, more akin to words than text.

Consider two strings of letters: "flirble" and "gwdihw." One has broadly agreed meaning, while the other is just an assemblage of letters but, for most readers, further information will be needed to know which is which. Meaning for whom? And in what language? Meaning depends on context. In Welsh, the letters "gwdihw" sound like an owl (hoo-dee-hoo); they represent that word. Gwdihw (in Welsh) and owl (in English) both convey similar meaning: a group of large-eyed nocturnal birds, Order Strigiformes. Flirble, meanwhile, follows standard rules for a word in English but has no broadly accepted meaning. A friend found this letter combination and uses it for his webserver. Strictly speaking, it is a word and conveys meaning to a small group of people who use the server. For most of the world, it is just a string of letters.

Strings of letters cannot hold meaning by themselves. They require a population of writers and readers who create (and sustain) that meaning through language. A word requires the agency of a writer—to encode meaning and set it down in text—and the agency of a reader—to observe the text and recover the meaning.

Like words, genes are inherently teleological. They pass form from parent to offspring. The genes only function properly when they do so accurately—when they pass on adaptive traits. [1] Words have a proper end: accurate communication between people. Genes also have a proper end: accurate transmission of traits. Biologists invariably use the language of agency and semantics: to send and receive signals, to encode and decode messages, to convey and interpret meaning. But what agents communicate using genes?

The gene mechanism was slowly worked out. The meaning of a gene is a trait or phenotype, nominally a protein. It is written in a chromosome as a sequence of nucleotides using a standard code. Chromosomes are copied and passed from generation to generation (and horizontally). Nucleotides and trinucleotide codons are the letters that make up genetic words.

Gene teleology remained obscure. Who writes and reads genes? The proximal efficient cause of a gene is a DNA polymerase. But that does not explain "by whom," "for whom," or "for the sake of which" the gene exists. A pencil may be the proximal efficient cause of a string of letters; it inscribes them. But writing words requires a more distal agent—a hand and a mind to form the words. A pencil cannot distinguish between gwdihw and flirble and a polymerase cannot distinguish between sense and non-sense sequences. The quest for a meaningful difference maker requires additional agency related to the ends and interests of writing. It requires power, form, and function beyond the simple mechanics of the biochemistry.

Early gene concepts provided a simple unit for vital forms without resolving questions of power and function. Localized to nucleic acids, they

---

[1] Perfect fidelity is not required for the transmission of information. Words allow people to communicate ideas, albeit imperfectly. Genes allow organisms to pass on adaptive traits, albeit imperfectly. Godfrey-Smith (2009) explores the effects of variation in fidelity when modeling inheritance. Increasing entropy entails, and natural selection requires, some noise in the signal. By "accurate" transmission, I mean sufficiently accurate for communication or for evolution by natural selection.

could be physical molecules as well as formal particles, but they still required biological agents to attribute meaning to the molecules, to inform the matter. Genes required something extra to take on the mantle of biological agency for themselves.

Genes as Agents

The agency of genes began with agential metaphors for molecules. Biochemists sought autocatalytic and autopoietic entities capable of being their own efficient cause. Like Aristotle's souls, they could cause themselves and provide a recursive explanation for the dynamism of life. Placing such power in a physical body (as Aristotle did with vegetable souls) should not obscure the fact that the power itself remains uniquely biotic. It does not address how biotic powers arise from abiotic physics. [2]

As noted in Chaps. 1 and 2, the "gene" that has agency is not a single polynucleotide. It is a collection of polynucleotides that share a sequence and a role in the evolution of a population. Philosophers distinguish between type and token. Gene tokens correspond to discrete physical molecules, but gene types reflect abstractions used to describe a population. Power and function were attributed to the type. [3] Polynucleotides are passive material particles. Genes, as agents, are abstract formal particles. [4] This is not to say that they are incompatible with natural science, only that their compatibility came from a rejection of prospect and not from an embrace of strict mechanism or strict physicalism. They retain that strange intermediacy by which vital particles empower and vital forms inform living processes. Like vegetable souls and their successors, they embed teleology in material processes.

---

[2] Aristotle was satisfied with an infinite regress of self-causing souls. Early twentieth-century biologists still sought to solve the problem historically, but autocatalysis and autopoiesis do not in themselves accomplish this. They remain open to vitalist interpretations without a parallel theory of abiogenesis.

[3] Neander (1991, p. 460): "Selection is always of types, not tokens. So, function attributions belong primarily to types and only derivatively to tokens." Haig (2012, p. 476): "the actions of a strategic gene are always judged by the criterion of the collective self-interest of its tokens considered as a group."

[4] To be clear, they are particles because they are discrete units of inheritance. They are informational parts, whether or not they are material parts. They are immaterial because they are abstractions, whether considered as ideal forms or nominalist categories.

Biological Interest

Organisms and groups were taken as the proper loci for ends well into the twentieth century. Organisms had interests and organisms coded information in genes for the sake of those interests. Most biologists focused on selfish explanations—individual organisms acting in their own interests. Alternatively, some biologists argued that species had interests and that genes served those ends. They appealed to the collective interest of groups, emphasizing the "function" of an individual within a population. They felt this explained altruistic behaviors—individuals acting for the benefit of others against their own interests.

Prominent figures in the Modern Synthesis dismissed group selection, arguing that it was mathematically impossible, due to concurrent selection at the individual level (Fisher 1958; Williams 1966). Group-level selection for a trait might promote one group over another, but organism-level selection against the same trait would eventually remove it from the group. In the long term, individual interests would win out. Current theories of group selection attempt to circumvent this problem with appeals to kin selection (Maynard Smith 1964; Sober and Wilson 1999) or reciprocity (Trivers 1971), which may be viewed as egoism at a larger scale. In the late nineteenth and early twentieth centuries, however, group selection was often associated with progressive theories of evolution, particularly ideas about the predictable development of human societies; individual altruism was explained by the health of the group.

In any case, selfishness and altruism are both inherently teleological. They require identification of interests, whose interest they are, and who has the power to act on them. They follow closely a conceptual framework established with human agents in mind. [5] Thus, arguments about levels of

---

[5] The individual versus group selection debate in biology mirrored a contemporary discussion in sociology. Is it better to focus on individual or group interests? Auguste Comte (1798–1857) and Émile Durkheim (1858–1917), pioneers of sociology, introduced a language for understanding human agency and interests. The term "egoism" arose in the eighteenth century to describe "self-interest." Comte championed the term "altruism" to describe behavior for the benefit of others. Durkheim promoted "functionalism," a theory that focused on how individual human actions and morality serve societal interests. His position was contrasted with that of Max Weber (1864–1920), who focused on the interests of individuals. Comte and Durkheim saw themselves as applying Darwinian evolution to human societies and (following Darwin's *Descent of Man*) proposed progressive theories of social evolution.

selection were arguments about ends, agency, and power—albeit in a modified form.

In pre-Enlightenment biology, an organism could be said to have a "function" because it participated in something greater than itself. The cosmic organism had an end (e.g., the love of God for Thomas Aquinas) and that end informed individual ends. Enlightenment dualism and the loss of the cosmic organism called into question the whole idea of organismic function. [6] Modern biologists must ask what it means to have function-like ends at the level of a "whole" organism.

Mayr (1961, p. 134) argued compellingly that Aristotelian teleology could be rehabilitated only in part. Mayr ruled out universal ends, including a purpose for nature or natural selection, which seem incompatible with natural science. What is the largest unit for biological ends? And do the ends of the largest unit (i.e., individual fitness or health) differ from the ends of parts (i.e., function)? These remain open questions in biology.

Genetic Interest

W. D. Hamilton (1963, p. 354) provided a key insight that reframed questions of scale. "Despite the principle of 'survival of the fittest' the ultimate criterion which determines whether G will spread is not whether the behavior is to the benefit of the behaver but whether it is to the benefit of the gene G." This, perhaps for the first time, made the gene the "for whom?" of bioteleology. Genetic ends became important.

Hamilton described "inclusive fitness," wherein selection calculations include both personal advantage and advantage to relatives in the degree to which they are related. If an action provides twice the benefit to full sibs (50% genetic identity) as it costs the organism (100% genetic identity), then the genes break even. More benefit to sibs and a gene can be selected for, despite an appearance of altruism. The organism is self-sacrificing, but the gene is selfish, and egoism wins the day.

A gene token, a single polynucleotide could not do this work. The benefit accrues to the gene type. Haig (2012) argues for a *strategic gene*, a set of gene tokens, usefully described as having causal power ("genes can be considered agents that benefit from the phenotypes they cause," p. 466), information content ("gene sequences come to embody and represent

---

[6] For a brief history of the world as organism and the replacement of world as machine metaphors, see Ruse (2010). For a deeper dive, see Lovejoy (1971).

'information' about what works in the environment," p. 464), and function ("Teleological language is appropriate when referring to the functions of genes because functions are final causes. They are both causes of a gene's persistence and effects of the gene," p. 465). Significantly, Haig emphasizes the causal role played by the environment and the importance of matching gene environment (including conditions inside the cell and organism) to gene action and fitness. [7] Nonetheless, he unambiguously attributes agency to the strategic gene and specifically mentions power, form, and function.

Power, form, and function have not been removed from modern biology; they have been recast as properties of genes and populations. Genes retain the agential language and conceptual framing of vital particles, including life-specific dynamism, selfishness, and recursive causation. The processes of natural selection, meanwhile—environmental interaction with populations—retain the language of life-specific "forces." Some of the most heated controversies in evolutionary theory arose on precisely these issues. How warranted is the use of agential language when speaking of natural processes?

Thus, there is a continuity between vegetable soul and gene concepts—a lineage of biological actors. What conceptual adaptations have shaped the changing population of concepts to fit them to natural science?

A NEW TELEOLOGY

Genes continue a tradition of teleological appeals in biology. They inherited power, form, and function from vegetable souls by way of vital particles and vital forms. The environment remains an active player in natural selection reflecting some of the "Lamarckian" teleology of vital forces. Claims that biological teleology has been naturalized (Sober 1993), mechanized (Ruse 2016), or historicized (Griffiths 2018) describe an important development in biology, but they can be misleading when they conflate multiple changes, which occurred at different points in the history of biological actor concepts. Evolutionary theory has benefitted from three distinct constraints on bioteleology. They are covered briefly here, more fully in Chap. 7.

First, Descartes and Bacon rejected transcendent or unnatural final causes. If such ends exist beyond nature, they also exist beyond the reach

---

[7] Haig (2020) expands on the theory.

of natural science. Coherent, but not dogmatic naturalism, won the day by allowing both epigenetic power and preformationist information to reside in material bodies. The bounds of "nature" have changed over the centuries, but the prohibition consistently does work. With Mill, we can recognize the importance of using the boundary consistently. For genes to be natural biological agents, they must follow regular laws that accord with the physics and chemistry of nucleic acids.

Second, Darwin provided evidence against progressive evolution. Starting in the eighteenth century, biologists began to accept that species change through time. Many embraced immanent, natural and historical progress. In other words, they believed this change reliably improved species and individuals in the long term. Darwin's branching tree of life refuted this perspective. It showed that change was not always for the better. Biological actors, insofar as they pull, do not always pull together. Life at large diversifies. From the perspective of biology as natural science, the power, form, and function of living beings serve no universal common end. Neither cosmic soul nor cosmic organism make sense in this context. Teleology should always be approached at the level of individuals acting within a specific context of population and environment. The individuals have changed to include genes, but the principle has not. Biological actors are contingent, local, and integrated with their surroundings. [8] Genes as agents, units of selection, and as beneficiaries of teleology are group-level phenomena and not sub-organismal.

Third, evolution by natural selection can provide power, form, and function without prospect. Biologists need not invoke "goal-seeking" by God, nature, or individuals. [9] This does not exclude prospective environments, as in artificial selection, nor prospective individuals, such as humans.

---

[8] Talk of evolutionary "trends" has been contentious, largely due to association with the three forms of teleology discussed here and in Chap. 7. They must be natural, context-specific, and blind. Modern definitions of a trend in biology emphasize consistent change of some trait at the population level. "Trends are persistent statistical tendencies in some state variable(s) in an evolutionary time series" (McKinney 1990, p. 55). This definition emphasizes that the change occurs stochastically at the group level. Questions of agency have been sidestepped by speaking of collectives changing collectively. Biological trends occur locally, and local conditions are essential to understand them, or even to meaningfully differentiate them from background noise (Gregory 2008). Nothing prevents global, that is Earth-wide, trends from occurring or being observed in biology so long as the relevant environmental factors can be identified. For more on trends, see Mix (2022).

[9] For examples of biologists making such appeals, see Chap. 5. Haller invoked God. Wallace invoked Spirit. Lamarck invoked a natural path to perfection.

Rather, it emphasizes the role of natural selection in providing power, form, and function even when prospect is absent. Fisher (1934) and Mayr (1961) emphasized that evolution can produce prospective agents but does not require them. Blind chance and mathematical necessity suffice.

The Modern Synthesis introduced models of biological action that retained biology-specific power but eliminated prospect. Genes regulate, but they do so blindly. Environments shape, but they have no vision. Common statements deny intent (Monod 1972; Nagel 1977; Mayr 1992; Walsh 2008; Larson et al. 2013). Other labels include purpose (Mayr 1961, 1992), goal direction (Simpson 1964), conscious design (Lennox 1993), representation (Reiss and Reiss 2005), and foresight (Larson et al. 2013). Power and prospect both raise difficult issues for philosophers, but prospect alone proved unacceptable to biologists.

The language of gene "selfishness" or "altruism" reflects a dramatic repurposing of those words. This has been justified by evolutionary theory, but evolutionary "altruism" must be recognized as profoundly different from sociological, psychological, and common sense "altruism."

*   *   *

Natural science and evolutionary thinking placed constraints on biology, selecting for biological actor concepts useful for empirical reasoning and prediction. Transcendence, progress, and prospect were excluded. Their reintroduction will be deleterious unless or until such a time as the conceptual environment changes. And yet, biological "agency" remains in gene concepts (with organisms and, possibly, groups) and in the "action" of the environment on populations.

The transcendent bioteleology of the Middle Ages is no longer tenable: interpretations of Aristotle that place final and formal causes in the mind of God. Vegetable souls make no sense after the Enlightenment. Neither can the strict mechanical naturalism of Descartes be maintained. Genes hover awkwardly in the middle ground between immaterial agents and passive matter. They are, in their way, active though careful attention will be required to see how this may be. Since the Enlightenment, teleology has become more natural and nature more agential to accommodate the strange dynamic consistency of living things.

The next chapter looks more closely at naturalism, contingency, and blindness as modern questions. The key issues of final causes which troubled Ancient, Medieval, and Enlightenment thinkers continue to challenge theoretical biologists and philosophers of biology. Many have proposed breaching the bounds set by previous generations of biologists, but the three constraints remain adaptive and important for the success of evolutionary theory.

REFERENCES

Fisher, Ronald A. "Indeterminism and Natural Selection." *Philosophy of Science* 1, no. 1 (1934): 99–117.

Fisher, Ronald A. *The Genetical Theory of Natural Selection,* 2nd ed. New York: Dover, 1958.

Godfrey-Smith, Peter. *Darwinian Populations and Natural Selection.* Oxford: Oxford University Press, 2009.

Gregory, T. Ryan. "Evolutionary Trends." *Evolution: Education and Outreach* 1, no. 3 (2008): 259–273.

Griffiths, Paul. 2018. "Philosophy of Biology." In *Stanford Encyclopedia of Philosophy*, Spring 2018 ed. Stanford University, 1997–. https://plato.stanford.edu/entries/biology-philosophy/.

Haig, D. "The Strategic Gene." *Biology & Philosophy* 27, no. 4 (2012): 461–479.

Haig, David. *From Darwin to Derrida: Selfish Genes, Social Selves, and the Meanings of Life.* Cambridge, MA: MIT Press, 2020.

Hamilton, William D. "The Evolution of Altruistic Behavior." *American Naturalist* 97, no. 896 (1963): 354–356.

Larson, Gregor, Philip A. Stephens, Jamshid J. Tehrani, and Robert H. Layton. "Exapting Exaptation." *Trends in Ecology and Evolution* 28, no. 9 (2013): 497–498.

Lennox, James G. "Darwin *was* a Teleologist." *Biology and Philosophy* 8, no. 4 (1993): 409–421.

Lovejoy, Arthur O. *The Great Chain of Being: A Study of the History of an Idea.* Cambridge, MA: Harvard University Press, 1971.

Maynard Smith, John. "Group Selection and Kin Selection." *Nature* 201, no. 4924 (1964): 1145–1146.

Mayr, Ernst. "Cause and Effect in Biology." *Science* 134, no. 3489 (1961): 1501–1506.

Mayr, Ernst. "The Idea of Teleology." *Journal of the History of Ideas* 53, no. 1 (1992): 117–135.

McKinney, Michael L. "Classifying and Analysing Evolutionary Trends." In *Evolutionary Trends*, edited by Kenneth J. McNamara, 28–58. Tucson: University of Arizona Press, 1990.

Mix, Lucas J. "Distinguishing Biological Trends from Adaptation." *Philosophy, Theory, and Practice in Biology.* 14 (2022): 10.

Monod, Jacques. *Chance and Necessity*. New York: Vintage Books, 1972.

Nagel, Ernest. "Goal-Directed Processes in Biology." *Journal of Philosophy* 74, no. 5 (1977): 261–279.

Neander, Karen. "The Teleological Notion of 'Function.'" *Australasian Journal of Philosophy* 69, no. 4 (1991): 454–468.

Reiss, John H., and John O. Reiss. "Natural Selection and the Conditions for Existence: Representational vs. Conditional Teleology in Biological Explanation." *History and Philosophy of the Life Sciences* 27, no. 2 (2005): 249–280.

Ruse, Michael. "Evolutionary Biology and the Question of Teleology." *Studies in History and Philosophy of Science Part C: Studies in History and Philosophy of Biological and Biomedical Sciences* 58 (2016): 100–106.

Ruse, Michael. *Science and Spirituality: Making Room for Faith in the Age of Science*. New York: Cambridge University Press, 2010.

Simpson, George G. "The Nonprevalence of Humanoids." *Science* 143, no. 3608 (1964): 769–775.

Sober, Elliott. *Philosophy of Biology*. New York: Oxford University Press, 1993.

Sober, Elliott, and David S. Wilson. *Unto Others: The Evolution and Psychology of Unselfish Behavior*. Cambridge, MA: Harvard University Press, 1999.

Trivers, Robert. "The Evolution of Reciprocal Altruism." *Quarterly Review of Biology* 46 (1971): 35–57.

Walsh, Denis M. "Teleology." In *Oxford Handbook of Philosophy of Biology*, edited by Michael Ruse, 113–137. New York: Oxford University Press, 2008.

Williams, George C. *Adaptation and Natural Selection: A Critique of Some Current Evolutionary Thought*. Princeton, NJ: Princeton University Press, 1966.

CHAPTER 7

Can Teleology Be Saved? Three Constraints on Bioteleology

Abstract The dynamic regularities of living things still demand a language of ends and a theory of power, form, and function to go with it. Ancient and Medieval formulations of souls with final causes failed, but something new had to take their place. Modern biologists embraced methodological naturalism, local adaptation, and blind chance. Though they frequently overlap, they are conceptually and epistemically distinct, requiring distinct defenses. Each places a necessary, though not sufficient, constraint on teleology within modern biology. Methodological naturalism excludes a priori agents that do not act with lawful regularity. Local adaptation reflects an a posteriori discovery that evolution is neither progressive nor linear, though consistent changes may occur under consistent conditions. Blind chance describes the sufficiency of genes and natural selection to explain evolution without invoking prospect, will, or other mental teleologies. Modern discussions of bioteleology can be improved with a recognition that the three constraints represent a Venn diagram of overlapping debates, each with its own set of explanatory and historical issues.

Keywords Chance • Evolution • Modern synthesis • Naturalism • Progress • Teleology

L. J. Mix, *The End of Final Causes in Biology*,
https://doi.org/10.1007/978-3-031-14017-4_7

Centuries of discussion have only highlighted the challenges of speaking about teleology in biology. The dynamic regularities of living things still demand a language of ends and a theory of power, form, and function to go with it. Ancient and Medieval formulations of souls with final causes failed, but something new had to take their place. Chapter 5 covered a variety of biological agent concepts and Chap. 6 covered genes and outstanding questions of biological agency. This chapter looks more closely at the conceptual environment in which biological agent concepts and bioteleologies evolved. How have modern biologists limited acceptable forms of teleology?

Final causes attracted regular debate prior to Darwin (Sloan 2017; Mix 2018) and from Darwin until the early twentieth century (Ulett 2014; Bowler 2017). Despite continuing use of ends language and concepts, the terms "teleology" and "final cause" were broadly rejected in the mid-twentieth century. Biologists rejected specific kinds of bioteleology deemed to be unhelpful. The terms were rehabilitated in the 1970s and 1980s along with a renewed appreciation for the original insights of Aristotle (Perlman 2004; Allen and Neal 2019). Biologists and philosophers in this period attempted to clarify and codify theories of function in a way that transparently avoided progressive evolution and supernatural intentional agents. And yet, bioteleology has a broader scope than just function. Modern biology includes "teleology" in a more general sense: it employs ends language in discussion of fitness, adaptation, and health. This discussion has produced three distinct constraints on bioteleology: methodological naturalism, local adaptation, and blind chance.

Each constraint reflects a specific historical context and specific style of argumentation. They are linked but each can only be properly understood on its own terms. Bacon and Descartes rejected final causes at the beginning of the Scientific Revolution. Their exclusion of supernatural teleology, specifically ideas in the mind of God, led to an a priori argument for methodological naturalism. Their work influenced Darwin in his treatment of design, function, and natural selection (Lennox 1993, 2010; Huxley 1874). Darwin also rejected "final causes," but his use of the term was not identical to that of Bacon and Descartes, nor was his concept of "nature." He, nonetheless, affirmed methodological naturalism and introduced a second constraint. Disproving the progressive teleology of Lamarck and other contemporaries, he replaced it with local adaptation. The Modern Synthesis added a third constraint: blind chance. Twentieth-century biologists rejected the prospective agency of orthogenesis and

"Lamarckism," replacing it with stochastic interactions between genes within populations in the context of a local environment.

Bioteleology remains an active area of research within the three constraints. Nor should my simple narrative of additional restraints be taken too linearly. Philosophers and biologists have questioned both the constraints themselves and how strictly they should be applied. Each of the three constraints has been reinterpreted over the centuries. And each must be defended on its own terms to keep biology moving forward.

Numerous biological questions cannot be meaningfully answered until some consensus is reached on the proper use of ends language. Should we treat biological ends as objective features of nature or only as useful metaphors (Papineau 2003; Depew 2010; Haig 2014)? How much work can we expect metaphors to do (Keller 2002; Midgley 1985: pp. 114, 143–154)? Do functional biologists (e.g., molecular biologists and medical researchers) and evolutionary biologists have different concepts of teleology (Mayr 1961; Haig 2013)? More concretely, biologists ask methodological questions about how we identify biological ends. How do we distinguish between functional and non-functional regions of a genome (e.g., Graur et al. 2013)? How do we distinguish between adaptive natural selection and random drift (e.g., Millstein 2002; Mutumi et al. 2017)?

Methodological naturalism, local adaptation, and blind chance all contribute to these discussions. Though they frequently overlap, they are conceptually and epistemically distinct, requiring distinct defenses. Each places a necessary, though not sufficient, constraint on teleology within modern biology.

Muddled "Teleology"

Multiple, conflicting definitions of teleology make it difficult to think critically and speak persuasively about biological agents. Public and academic discussions circle around a variety of issues and the term often carries negative, expressly unscientific, connotations.

The *Oxford English Dictionary* provides a clear example. The free web edition defines teleology as "The explanation of phenomena in terms of the purpose they serve *rather than* of the cause by which they arise."[1] The

---

[1] *Oxford Dictionaries, English*, s.v. "teleology." Accessed March 22, 2019. https://en.oxforddictionaries.com/. Italics added. This reflects Descartes' critique of Gassendi's use of final causes; see Osler (1994, 162).

only sub-entry refers to theology. Entry and sub-entry both suggest that teleology excludes natural causation and is incompatible with modern science. The full *OED* entry provides more information, starting with "The branch of knowledge or study dealing with ends or final causes."[2]

If biological ends act with and through biochemical mechanisms, they need not compete with familiar scientific causes. Recent scholarship suggests that Aristotle had just such a concordance in mind, though his Medieval interpreters did not (Chaps. 2 and 3; Gotthelf 1976; Johnson 2005; Mix 2016). Confusion is unsurprising when the formal and freely available works of the same authority provide different definitions. Many biologists, philosophers, and historians lump all teleology together and dismiss it as unscientific, disproved, and/or reformed. But these cannot all be the same "teleology."[3]

Categories of Teleology

Allen and Neal (2019) discuss a variety of modern attempts to reconcile teleology with biology. They identify five common complaints about teleology as unscientific. It has been rejected as vitalist, anachronistic (invoking backward causation), mentalistic, non-mechanistic, and

---

[2] *Oxford English Dictionary*, 3rd ed. (2010), s.v. "teleology, n."

[3] Ernst Mayr (1961) and Jacques Monod (1972) famously argued for licit biological ends but renamed them. *Teleonomy*, they said, describes programmed behavior. [Mayr (1961) borrowed the word from Colin Pittendrigh (1958) but adjusted the meaning. Pittendrigh rejected "Aristotelian teleology" writ large. Mayr thought Aristotle had two teleologies, only one of which was useful.] Beyond this narrow range, *teleology* reflects unscientific thinking. Walsh (2008) tells us that "received opinion" unequivocally rejects teleology. He then argues for a newly articulated, "naturalistically acceptable" teleology. Ruse (2016) declares that "Evolutionists want teleology but they cannot have it." Sober (1993: pp. 82–87) provides a "naturalized teleology" compatible with modern physics and biology. Griffiths (2018) reports that teleology has been historicized, and thus made acceptable, by Ruth Millikan and Karen Neander. One might add the naturalized teleologies of Robert Cummins, Carl Hempel, Alvaro Moreno, Ernest Nagel, Norbert Weiner, William Wimsatt, Larry Wright, and many others.

non-empirical—all concepts as controversial as teleology itself.[4] One obscure problem has been yoked to five more. The demarcation of science and empirical reasoning continue to trouble philosophers (Hansson 2017). Despite this breadth, most discussions of teleology are dualistic, opposing teleology and modern science.[5] While informative, they can obscure the variety of ways that teleology can conflict with modern biology.

Several philosophers have explored the range of theories available. Some provide simple dichotomies between causal and relational accounts of function.[6] Others present complex taxonomies with diverse species of teleology.[7] Such distinctions can be valuable for teasing out the different aspects of teleology, but they can constrict discussion in two ways. First, most focus exclusively on function. Second, by separating naturalist from non-naturalist teleologies at the base of the tree, they preempt substantive discussion about the meaning of "natural," the naturalness of minds, and the alternative ways of being unnatural.

"The notion of function is not all there is to teleology, although it is sometimes treated as though it were" (Wright 1973, p. 139). Several prominent figures in this discussion emphasize the role of teleology

---

[4] Allen and Neal find the first four in Mayr (1988); a fifth comes from Allen and Bekoff (1995). Rey (2000) and Canguilhem (2008) argue that "vitalism" is used uncritically as an insult, a way of labeling a theory unscientific without committing to the bounds of science. The limits of "mechanical philosophy" have been controversial from its foundations (Osler 1994, 2010). The proper understanding of minds remains a hot topic, exemplified in discussions of will (e.g., O'Connor 2009; Dennett 2015) and the "mental" life of plants. Chamovitz (2012) and Wöhlleben (2016) summarize recent scientific work on plant faculties similar to sensation and memory. Miller (2002), Hall (2011), Nealon (2015), and Irigaray and Marder (2016) trace historical and social issues related to treating plants as agents with their own subjective ends.

[5] Popular dualisms include backward causation versus etiological function (Neander 1991), vitalist versus physicalist approaches (Sober 1993), skyhook versus crane (Dennett 1995), and conditional versus representational teleology (Reiss and Reiss 2005). Allen and Neal (2019, sec. 2) distinguish "teleomentalist" and "teleonaturalist" accounts. These pairs correspond to a colloquial division between scientifically taboo "purpose" and scientifically respectable "function." *Wikipedia*, s.v. "function (biology)." Accessed August 29, 2019. https://en.wikipedia.org/wiki/Function_(biology).

[6] Godfrey-Smith (1993) distinguishes the etiological functions of Larry Wright (1973) from the systemic functions of Robert Cummins (1975). Walsh and Ariew (1996) similarly divide evolutionary E-functions from capacity-based C-functions. Contrary to Godfrey-Smith (1993), they link all etiological accounts to evolution.

[7] For example, Allen and Bekoff (1995), Perlman (2004), and Allen and Neal (2019).

beyond function and the importance of thinking more broadly (e.g., Mayr 1961; Neander 1991). Previous chapters have looked at power and form. Currently popular definitions of life depend critically on understanding the unique dynamism of Darwinian populations and homeostatic regulators (Godfrey-Smith 2009; Mix 2015). No less than Aristotelian theories, they require attention to how the power to do work comes to be caught up in these systems. Teleology becomes important for astrobiology and origin-of-life studies that address how biospheres arise. Normative form and function will also be key to understanding medicine, bioethics, and the ends we attribute to whole organisms. The function literature addresses important questions in philosophy of biology, but a broader understanding of teleology is still required.

Turning to "nature," historical studies have revealed a wide range of teleological theories in biology, some more and some less compatible with modern science (Lenoir 1982; Lennox 2010; Sloan 2017; Mix 2018). Modern philosophers remain divided. Papineau (2016) states that naturalism, while popular, "is not a particularly informative term." It may be ontological or methodological, reductive or non-reductive. It may or may not include consciousness, moral facts, and mathematical facts. As the last two chapters showed, biologists can fall into competing camps with both sides claiming the naturalist banner.

Teleology, *as ends language in some form*, is necessary. And yet, many specific teleologies have been justly rejected as unscientific, disproved, or naturalized. Articulating three distinct constraints allows biologists to see how they work and how they may—or may not—be linked.

METHODOLOGICAL NATURALISM

Robert Pennock (1996a, b) defends a position called methodological naturalism. He argues for a "lawful regularity" at the heart of the natural science worldview. The subjects of astronomy, biology, chemistry, and so on are bound by fixed laws, knowable to the scientists who study them. Natural laws bind natural objects and constrain natural events. Significantly, Pennock's naturalism is methodological. He does not presume that all objects and events are natural, a view he calls ontological naturalism. Instead, he commits to the usefulness of natural science, while remaining agnostic about other forms of knowing. Critically, the existence of a border is more important than its precise location.

Pennock (1996a, p. 551) argues against supernatural explanations in science: "to say that some power is supernatural is, by definition, to say that it can violate natural laws." Supernatural causes, once accepted, fail to add to scientific knowledge. Unbounded by lawful regularity, they remain unpredictable.

Many readers will want a stronger version of naturalism, one that explicitly excludes souls, spirits, and vital forces (or a similar list of entities). Such embellishments, however, can obscure a key insight of the Enlightenment: the comprehensibility of the natural world. Bacon and Descartes championed natural science, but neither placed the borders of nature in the same place as modern scientists. Thus, it is not obvious that their rejections of final causes should entail a similar rejection within modern biology. Over the past few centuries, the borders have shifted dramatically. Nature now excludes animal spirits, vital forces, and infinitely divisible matter. It includes vacuum, quantum indeterminacy, and probabilistic processes. Prominent biologists of the nineteenth century embraced theories now considered unacceptably and unnaturally teleological (e.g., Oken's *galvanism* or Lamarck's path to perfection). Future philosophers may find current biology equally untenable.

None of this detracts from the value of embracing lawful regularity discoverable in the context of natural science. Lawful regularity alone will be sufficient to reject many popular theories. In committing to biology as natural science, biologists chose to limit themselves to lawfully regular explanations. The laws themselves remain open for debate, but some laws must exist and be consistently obeyed.

Missing the Point

Methodological naturalism is a theory about how we create knowledge, specifically knowledge as natural science. Authors who introduce natural scientific evidence against methodological naturalism miss a crucial distinction between the a priori epistemological issue—science or nonscience—and the scientific one—the quality of evidence.

Scientific evidence presupposes lawful regularity. Calls for "unnatural" actors or events in "natural" explanation are self-contradictory. If an event is regular in one way but not another (e.g., probabilistic laws where single events cannot be predicted but aggregate outcomes can), then it is the regularity that is relevant for scientific explanation. The irregularity is interesting. It can prompt research. It may reflect causal ignorance (cause unknown) or fundamental chance (cause unknowable, i.e., *Tychism*). But the regularity makes it natural or scientific (as natural science).

This is a common error in discussions of "random" evolution. Positive statements about stochastic variation are scientific; they describe regular changes in populations. As we will see below in the section on blind chance, positive statements can also be made about the sufficiency of explanations with minimal regularity. Physicists can make dramatic predictions based on quantum mechanics, despite a lack of knowledge about individual particles. Knowledge about large populations suffice. Similarly, biologists can make predictions about populations based on population dynamics and environmental constraints, despite a lack of knowledge about the outcomes of individual organisms.

Negative statements about "random" evolution excluding intentional agency (divine or human) are not scientific; they impose irregularity on a system when regularity may at some point be discovered. Such claims make natural selection and artificial selection mutually exclusive. Human agency—in artificial selection—would have to be fundamentally unpredictable and irredeemably unnatural. Most biologists would, I suspect, be unwilling to grant humans this status. The "randomness" of evolutionary theory does work as a positive statement about regularity and predictability; it does not exclude agents or proto-agents.[8]

Conversely, appeals to inherently unnatural or unpredictable agents move bioteleology beyond the realm of natural science. Having denied regularity, no amount of data will restore the naturalness of the explanation. Biologists (as natural scientists) affirm the sufficiency of lawfully regular explanations for a broad range of phenomena.

Koperski (2008, p. 440) argues that, if naturalism is truly methodological, scientists must be open to finding evidence for unnatural actors. "It borders on academic incompetence to pretend that science has strict boundaries and then gerrymander those boundaries to keep out the riffraff." And yet, he does precisely that. The boundaries of natural science are methodological. Nature cannot be gerrymandered to include unnatural agents. To deny the inclusion of designers in nature, one must deny proof of designers from nature.

---

[8] This conflation of different meanings of random has, additionally, a negative impact on science outreach when it gets caught in religious questions about divine action (Mix and Masel 2014).

The advisability of natural science and the sufficiency of evolution within natural science are distinct questions. The first must be argued a priori; the second a posteriori. Adaptation remains the preeminent theory of biological change within the bounds of methodological naturalism.

A new epistemology may be desirable, but modern biology (constrained by the naturalism of Bacon and the evolution of Darwin) cannot provide evidence for it. In the words of Pennock (1996a, p. 557), "negative arguments against evolutionary processes are irrelevant to the key question." Unnatural teleologies cannot be meaningfully tested scientifically. This in no way excludes natural teleologies, which must be amenable to empirical proof or disproof.

Creationism Versus Evolution

To the extent that God behaves with lawful regularity, God may be usefully described as natural. After the Enlightenment, most Christian, Muslim, and Jewish theologians have rejected this position. They say that God's regularity is willfully chosen, not lawfully obeyed. No description of the law can resolve ambiguity about whether it will be obeyed. Teleology is unscientific when it requires unnatural elements. Scott (2009, p. 19) makes the point that natural science cannot test for a Creator, because creation cannot be held constant. All is created or nothing is. Sober (2008) provides a more extensive analysis, showing that unnatural design hypotheses add no predictive power to biology. Most modern arguments for Creationism and Intelligent Design can be excluded from biology with methodological naturalism alone.

LOCAL ADAPTATION

Lamarck and Darwin both accepted the methodological naturalism of Descartes and Bacon, but still thought teleologically. Evolution requires an end, namely, adaptation. They disagreed about where adaptation led. Lamarck saw a fixed trajectory, while Darwin saw populations drawn to changing local optima. This move from progressive evolution to branching evolution marks an important innovation and a second constraint on bioteleology. Biological ends are always local; they arise in response to specific populations, in specific environments, over specific timeframes.

Darwin accepted the a priori restriction on natural science and added a second constraint to bioteleology. Most earlier evolutionary theories involved progress along a fixed, upward path (Ruse 1996; Sloan 2017). Enlightenment thinkers embraced the ladder of nature (*Scala Naturae* or great chain of being) connecting lower and higher life (Lovejoy 1971). Lower life forms, now called bacteria, were seen as small, simple, mechanical, and worth little more than the mud in which they grew—literally, pond scum. Higher life forms, typified by humans, were large, complex, free, and dignified. All the steps between were filled with their own forms of life, links in an eternal chain. Early theories of evolution set the chain in motion, pulling species upward, but most biologists still viewed it as a single ascent from simple to complex (Mix 2018, pp. 191–196).[9]

Darwin disproved progressive teleology. As discussed in Chaps. 4 and 6, Darwin provided arguments against progressive evolution as a general rule. Evolution can lead to complexity, but it can also lead to simplicity. It can produce beauty, but also ugliness. He described an evolutionary ascent for humans, which has led to great debate about his teleology. But two points are critical in understanding his role in constraining bioteleology. First, his arguments (both for and against evolutionary progress) were natural science arguments, distinguishing this from the a priori rejection of "final causes." Second, he provided a way of thinking about bioteleology as local adaptation rather than absolute progress.

In *The Origin of Species*, Darwin concisely and systematically presented evidence against two ideas popular at the time. First, he showed that species diversify through time. They can become large, complex, and free in different ways. An orchid may be as advanced as a human, but advanced along a different path. Second, he showed that some paths lead to less size, complexity, and freedom. At the universal scale (all life) there is no progress. The ladder of nature was replaced by a tree, whose branches grow outward, even downward, as well as upward. With natural selection, Darwin argued that species adapt to local environments and not to some universal optimum.

Darwin did draw a connection between progress, providence, and design. Were it obvious that all variation and selection improved the natural world, one might attempt an end run around methodological naturalism. Long-term trends in biology, proved empirically, could be aligned with moral intuition, reinforcing both and supporting the idea of directed

---

[9] Lamarck (1914, pp. 56–61) referred to a path to perfection. Locke (1824, pp. 483–484) and Buffon (1797, pp. 255–272) provide contemporary parallels. Darwin rejected such theories, even when considered as natural (Ruse 1996, pp. 147–150).

evolution. Asa Gray hoped to leave the door open for just such an argument. Darwin emphasized the role of chance in natural selection, leading to a widening gap between the two (Lennox 2010).

All three approaches were teleological: Gray's theistic evolution, Lamarck's progressive evolution, and Darwin's branching adaptation. In Darwinian evolution, "Selection explanations are inherently teleological, in the sense that a value consequence (Darwin most often uses the term 'advantage') of a trait explains its increase, or presence, in a population" (Lennox 1993, p. 410). The key to Darwinian advantage, however, lies in its relativity. A trait may be advantageous in one context but not in another, against one competitor but not against another, over a short period but not over a long one. Evolution fits organism to population and habitat.

Lamarck contributed to the theory of evolution by highlighting the role of the environment and organism-habitat interactions, but he thought these forces led to absolute improvement. Darwin retained the interactions, but rejected the linear, positive trajectory, producing a theory of local—and only local—adaptation.

The Strong Anthropic Principle

In modern biology, evolution remains intractably local, but attempts to reintroduce the path to perfection appear regularly. The clearest examples relate to intelligence, often presented as an inevitable and irreversible adaptation. Intuitive appeals leverage common optimism about the quality, extent, and future opportunities for human intelligence, but it is not clear that intelligence is monotonically linked to success or that this link will continue indefinitely.[10]

Proponents of the Anthropic Principle claim broad knowledge about the future trajectory of life and intelligence. As with many theories, weak and strong versions exist (Mix 2009, pp. 58–66). The weak version is unambiguous and unambitious; the strong version is more interesting, but not analytically rigorous. The strong version of the Anthropic Principle involves naturalistic progressive teleology. A commitment to local adaptation allows us to reject it.

---

[10]Kurt Vonnegut's 1985 novel *Galápagos* makes this point in an engaging way. Koalas provide the classic example for adaptation that leads away from intelligence. Obligatory specialization in diet (low nutrient eucalyptus leaves) corresponds with small, smooth, metabolically inexpensive brains (Shipley et al. 2009).

Physicists John Barrow and Frank Tipler (1986) defend bioteleology in *The Anthropic Cosmological Principle*. Their weak anthropic principle (WAP) describes the selection effect brought about by humanity. An observed universe must have a kind of life that observes. (They take for granted that observers are living.) WAP is logically necessary and has no teleological content.

Their strong anthropic principle (SAP) is more ambitious. "The Universe must have those properties which allow life to develop within it at some stage in its history" (p. 21). Barrow and Tipler describe SAP as far more speculative and explore three different possible interpretations (pp. 21–23). They label the first "teleological" and associate it with Fred Hoyle's argument from stellar nucleosynthesis to cosmic fine-tuning. They find this unscientific, as it is impossible to prove or disprove the existence of a designer outside the universe. Barrow and Tipler's teleological SAP violates methodological naturalism.

They remain open, however, to natural ends, as becomes clear in their discussion of Teilhard de Chardin. "Although his specific cosmological model failed to correspond to reality, it is by no means impossible to construct a testable theory of a progressive cosmos which is roughly analogous to the Teilhardian theory" (127). They refer to it as an "indeterminate natural teleology." We can discern broad trends, but not specific outcomes. Nonetheless, they argue for a universal trend in the trajectory of life, one that leads to intelligent observers.

Barrow and Tipler also suggest a final anthropic principle (FAP) along these lines. "Intelligent information-processing must come into existence in the Universe, and, once it comes into existence, it will never die out" (p. 23). This approaches Lamarck's path to perfection as it suggests a non-reversible process free from local contingency which guarantees a particular, desirable end-state.

Local adaptation makes it clear that the SAP and the FAP fall outside the bounds of modern biology. Bioteleology must be local, reflecting a specific population in specific surroundings over a specific period of time. As noted by Mayr (1961) cosmic teleology is not allowed. If life can only be found in one type of habitat—perhaps, within the temperature range of liquid water—then life will always adapt to those circumstances. There may be common traits among localities. Thus, we can move beyond the WAP, but only when we can establish the similarity of habitats. Intelligence is a high-cost trait that is clearly not adaptive in some circumstances (e.g., stable low-risk, high-resource environments). Therefore, it is neither inevitable nor irreversible.

Is Progress Possible?

Local adaptation rules out the subsidiary-function teleology of pre-Darwinian evolutionary theory. No cosmic organism lends its telos to genes and organisms. Fitness and health cannot be organic function. Biological actors have ends of their own.

Progress has been ruled out as a normative claim about the trajectory of universe. If God has desires for nature, scientists lack reliable access. Insofar as humans have desires for nature, natural selection does not respect them. Progress has also been ruled out as an appeal to the ends of life-at-large, even if such "life" does not pervade the universe. Such arguments were common in nineteenth-century biology. Darwin showed that they were inconsistent with observation and Mayr ruled them out as the unacceptable portion of Aristotelian teleology.

I use the term "progress" to indicate a value judgment: movement toward a better outcome or a desired destination. I believe this to be the primary use of the term, both in public and in academic discussions of teleology. Normative and subsidiary teleologies of progress remain popular in folk biology and, as Barrow and Tipler demonstrate, non-biological academic discussions.

There remains the possibility of "progress" as value-neutral long-term trends in biology. Several authors have defined evolutionary progress in this way (Rosenberg and McShea 2007, pp. 139–168; Desmond 2021; DeCesare 2022). Given the popularity of value-laden teleologies and their extensive history in biology, I believe this use of "progress" is both misleading and potentially dangerous.

This is not to deny the existence of long-term evolutionary trends which may be discerned empirically (with great caution, see Mix 2022). Biological actors have consistently been defined teleologically in a way that suggests directionality. Aristotle's process of nutrition—form informing matter to make non-self into self—suggests movement toward the organization (that is the organism-ization) of all matter in the universe. This could be measured in terms of total biomass or copy number (for a particular form, such as a genome).

Unfortunately, the two measures need not agree. Consider life-history strategies for example. One can ask whether a million-cell organism is more advanced than a million-cell colony or a million dispersed clones. Natural selection appears to pursue multiple strategies. More problematically, once a biological actor has achieved a certain level of success, component parts begin to defect and become biological agents in their own rights (e.g., selfish genetic elements, cancer cells). Is any life more advanced than no life? Or can life only be said to progress relative to the interests of a particular actor? Worse yet, a universal organism would lack any external matter to incorporate and would, by definition, stop eating and cease to be alive. It may be contingently true that life expands over the full course of history, it cannot be necessarily true that life expands eternally.

Could there be a universal strategy, always adaptive for each (though not adaptive for all)? Arguments for convergent evolution hang on appeals to broadly adaptative traits. They remain useful to the exact extent that they track broadly common environments. Different circumstances usually lead to adaptation in different directions (Gregory 2008). A very small number of global (Earth-wide) adaptations have been tracked and attributed to consistent or consistently changing features of all Earth habitats.[11]

Barrow and Tipler may seem a trivial example. I have chosen it because it is clear and well known. Similar arguments remain common, however, in both popular and scientific contexts. Local adaptation rules out the value claims of Lamarckian progress. Universal trends might still be discovered, however, those trends most often proposed (e.g., toward intelligence and cooperation) have not borne out. Unlike supernatural teleology (ruled out of science a priori) and normative progress, long-term value-neutral trends remain open to empirical proof and disproof.

BLIND CHANCE

A third constraint on teleology fascinated biologists in the late nineteenth and early twentieth century and has come to dominate recent discussions—prospect or forward-looking agency. Methodological naturalism excludes explicitly unnatural actors. Local adaptation excludes universal progress. Many biologists attempted to explain biological power, information, and function using natural, local drivers. Proponents of the Modern Synthesis showed that blind chance, and thus blind agents, were sufficient for evolutionary theory.

---

[11] At the level of orders, maximal body size increases across geological time (Smith et al. 2016). Recent research suggests global long-term trends in protein structure (James et al. 2021) and genetic complexity (Wolf et al. 2018). McShea (2016) argues for several trends in complexity at a more abstracted level.

It remains unclear whether human agency can be compellingly natural-ized—even at the methodological level. Descartes and O'Connor (2002, 2009) argue that it cannot; humans are inherently unnatural. Hobbes and Dennett (1991, 2015) argue that it can. Or, perhaps they would say we need a more natural, non-standard, model of agency. I shall avoid the question of whether human prospect—and thus artificial selection—is natural. Luckily, biologists need not address this issue. The question of biological causation can be asked in a general way that applies to all organisms, including mindless vegetables. We can bracket the special case of human agency and think more seriously about the teleology of trees, protists, bacteria, and so on.

The power or proto-agency present in all living things requires no consciousness, interiority, or mind.[12] It was this more general view of biological causation that Darwin so famously tied to adaptation and natural selection. Some organisms survive and reproduce more efficiently than others; their environment "selects" among them. If a trait helps an organism survive and reproduce, and if that trait can be passed to offspring, the trait will increase within the population. The power resides jointly in the selecting environment and the reproducing population. This was the teleology that Neander (1991) and Millikan (1984) historicized and Sober (1993, pp. 82–87) naturalized.

After the Modern Synthesis, biologists rejected supernatural and progressive theories, but they also excluded a third class of explanation. The explanations that remained depended upon blind chance. Mutation, migration, selection, and genetic drift—all operating blindly—were shown to be sufficient for most, perhaps all, questions of interest to biologists. Biologists were adamant that biological causes need no prospect, no view of the future. Neither individuals nor species, nor natural forces, nor nature itself, requires prospective imagination. Theories that did not live up to this standard were labeled "orthogenesis," if the prospect was innate

---

[12] Mix (2018, pp. 8–10) discusses "proto-agency" and other analogues of human agency consistent with nature, including "primitive agency" (Burge 2009; Jones 2017, p. xiv), "minimal agency" (Barandiaran et al. 2009), and "doings" and "happenings" (Nida-Rümelin 2007).

to an organism, or "Lamarckism," if it was external.[13] Different thinkers labeled the constraint differently, but they all emphasized knowledge (or imagination) about the future and some form of agency, both of which are associated with minds and contribute to the objectionable character of prospect.[14]

The innate drivers of orthogenesis morphed into blind genes and enzymes. The extrinsic drivers of "Lamarckism" were reformed into natural selection: chance interactions between genes and organisms within a population and within an environment. Local adaptation took on mathematical form. Prospect disappeared and other questions of agency were tabled, hidden somewhere above the level of the gene (merely a molecule) but below the level of genome (a collection of genes), organism (merely a body), or population (a collection of organisms). This allowed the analysis, above and below, to be naturalized (Keller 2002, pp. 123–132).

This is not to say that biologists were being dishonest, either methodologically or ontologically. Rather, they found tractable questions at multiple levels. Biology provides naturalized actors and naturalized teleology useful at the level of gene, genome, organism, and population. As with human agency, some will be satisfied that this is sufficient for understanding "life" (e.g., Dawkins 1996); others will not (e.g., Midgley 1985, 2014). In either case, biologists—for the most part—were satisfied with this as a pragmatic solution.[15]

---

[13] Recall that Lamarck (1914) was not a "Lamarckist." He favored innate drivers, *subtle fluids* (pp. 187–188) and *orgasms* (pp. 211–229), which he expressly named natural, though they possessed some form of agency. They were not passive, as required by mechanical descriptions of nature, such as that of Descartes (Osler 1994). His emphasis on the role of the environment in evolution led many to identify him with extrinsic drivers, particularly when contrasted with Darwinian natural selection (Bowler 2017).

[14] Various theorists denied intent (Larson et al. 2013; Mayr 1992; Monod 1972; Nagel 1977; Walsh 2008), purpose (Huxley 1948, pp. 412, 576; Mayr 1961, 1992), goal direction (Simpson 1964), conscious design (Lennox 1993), representation (Reiss and Reiss 2005), and foresight (Larson et al. 2013). Fisher (1934) and Mayr (1961) went so far as to claim that prospective agents do not produce evolution, but evolution does produce them.

[15] Gardner and Welch (2011) and Haig (2012, 2020) present two examples of solid evolutionary theory tackling agency and teleology with philosophical rigor. Thus, I can appreciate Midgley's critiques of reductionism, dualism, and normativity in selfish gene descriptions—along with, for example, Meyer (2016) and Schloss (2004)—while also finding her concern about "intentionality" and "agency" misplaced—along with, for example, Gardner and Welch (2011). The ontological and epistemological questions—how we organize and explain the universe—deserve discussion. But, once we commit to natural science as *one* way of knowing, these terms have been naturalized (when used carefully in the context of evolutionary theory).

By attending to why and how blind chance and evolution naturalize some teleology—namely adaptive function—we see more clearly why it cannot naturalize other teleologies—such as design and moral progress. Given the current view of nature in the sciences (methodological and frequently debated, but good enough to get on with), evolutionary theory has naturalized teleology. It could not naturalize God, or Spirit, or the other explicitly unnatural agents of Medieval and Romantic biology; these are excluded from science a priori. It did not rescue the linear progress favored by Enlightenment biologists. That was disproved by Darwin and subsequent discoveries in paleontology, microbiology, and taxonomy. Instead, evolutionary theory has naturalized the ends of adaptation, function, and health by yoking them to stochastic processes, local conditions, and a new kind of blind "agent," the gene.

*   *   *

Bioteleology can only be understood by considering three distinct constraints. Influential thinkers explored theories on both sides of each bulwark, alone and in combination. Nicolas Malebranche (2000) promoted supernatural teleology while rejecting both progress and natural prospect. Erasmus Darwin (1796) eschewed supernatural teleology but embraced progress through prospective *sensoria* in even the most basic organisms. Jean-Baptiste Lamarck (1914) and Julian Huxley (1948) defended progressive evolution while affirming blind naturalism. Henri Bergson (1998) promoted prospective agency in the *élan vital* while affirming natural contingency and open-ended evolution. Each contributed to bioteleology, and each enjoyed praise among contemporary biologists because of their defense of one or two of these constraints. Each has been labeled "vitalist" and inappropriately teleological because they did not observe all three.

Modern discussions of bioteleology can be improved with a recognition that the three constraints represent a Venn diagram of overlapping debates. Simpler schemes (such as the teleonaturalist/teleomentalist divide of Allen and Neal, 2019) prevent discussion of natural prospect. Not only was this category important historically, it will be critical for future discussions of biological agency. How does the proto-agency of bacteria operate blindly? How does it contribute to the full-fledged agency of humans? If the second arises from the first, at what point does it become prospective, mentalistic, and/or unnatural? Positions on bioteleology will affect how such questions are asked and answered.

Three kinds of teleology have been rejected over the past three centuries. Each concept was maladaptive in the context of modern biology. Methodological naturalism, local adaptation, and blind chance constrain the current population of theories because they improve the conceptual coherence and predictive power of evolutionary theory. Chapter 8 returns to the question of what may be said within these bounds and what ends bioteleology might serve moving forward. A pragmatic approach to genes that respects the reciprocal evolution of population and environment can allow us once again to understand life as form informing matter through nutrition, growth, and reproduction.

References

Allen, Colin, and Marc Bekoff. "Function, Natural Design, and Animal Behavior: Philosophical and Ethological Considerations." In *Perspectives in Ethology 11: Behavioral Design*, edited by Nicholas S. Thompson, 1–47. New York: Plenum Press, 1995.

Allen, Colin, and Jacob Neal. "Teleological Notions in Biology." In: *Stanford Encyclopedia of Philosophy*, Spring 2019 ed. Stanford University, 1997–. https://plato.stanford.edu/entries/teleology-biology/.

Barandiaran, Xabier E., Ezequiel Di Paolo, and Marieke Rohde. "Defining Agency: Individuality, Normativity, Asymmetry, and Spatio-Temporality in Action." *Adaptive Behavior* 17, no. 5 (2009): 367–386.

Barrow, John D., and Frank J. Tipler. *The Anthropic Cosmological Principle*. New York: Oxford University Press, 1986.

Bowler, Peter J. "Alternatives to Darwinism in the Early Twentieth Century." In *The Darwinian Tradition in Context*, edited by Richard G. Deslisle, 195–217. New York: Springer, 2017.

Buffon, Georges Louis Leclerc. *Buffon's Natural History*, vol. 2. Trans. J.S. Barr. London: H. D. Symonds, 1797.

Burge, Tyler. "Primitive Agency and Natural Norms." *Philosophy and Phenomenological Research* 79, no. 2 (2009): 251–278.

Canguilhem, Georges. *Knowledge of Life*. New York: Fordham University Press, 2008.

Chamovitz, Daniel. *What a Plant Knows: A Field Guide to the Senses*. London: Farrar, Straus and Giroux, 2012.

Cummins, Robert. "Functional Analysis." *Journal of Philosophy* 72, no. 20 (1975): 741–765.

Dawkins, Richard. *The Blind Watchmaker*. New York: W. W. Norton, 1996.

De Cesare, Silvia. "Values in Evolutionary Biology: A Comparison Between the Contemporary Debate on Organic Progress and Canguilhem's Biological Philosophy." *History and Philosophy of the Life Sciences* 44, no. 2 (2022): 1–20.

Dennett, Daniel C. *Consciousness Explained*. New York: Little, Brown and Co., 1991.

Dennett, Daniel C. *Darwin's Dangerous Idea*. New York: Simon and Schuster, 1995.

Dennett, Daniel C. *Elbow Room*. Boston: MIT Press, 2015.

Depew, David J. "Is Evolutionary Biology Infected with Invalid Teleological Reasoning?" *Philosophy and Theory in Biology* 2 (2010).

Desmond, Hugh. "The Selectionist Rationale for Evolutionary Progress." *Biology & Philosophy* 36, no. 3 (2021): 1–26.

Fisher, Ronald A. "Indeterminism and Natural Selection." *Philosophy of Science* 1, no. 1 (1934): 99–117.

Gardner, Andy, and John J. Welch. "A Formal Theory of the Selfish Gene." *Journal of Evolutionary Biology* 24, no. 8 (2011): 1801–1813.

Godfrey-Smith, Peter "Functions: Consensus Without Unity." *Pacific Philosophical Quarterly* 74, no. 3 (1993): 196–208.

Godfrey-Smith, Peter. *Darwinian Populations and Natural Selection*. Oxford: Oxford University Press, 2009.

Gotthelf, Allan S. "Aristotle's Conception of Final Causality." *Review of Metaphysics* 30 (1976): 226–254.

Graur, Dan, Yichen Zheng, Nicholas Price, Ricardo B. R. Azevedo, Rebecca A. Zufall, and Eran Elhaik. "On the Immortality of Television Sets: 'Function' in the Human Genome According to the Evolution-Free Gospel of ENCODE." *Genome Biology and Evolution* 5, no. 3 (2013): 578–590.

Gregory. T. Ryan. "Evolutionary Trends." *Evolution Education and Outreach* 1, no. 3 (2008): 259–273.

Griffiths, Paul E. "Philosophy of Biology." In *Stanford Encyclopedia of Philosophy*, Spring 2018 ed. https://plato.stanford.edu/archives/spr2018/entries/biology-philosophy/.

Haig, David A. "The Strategic Gene." *Biology and Philosophy* 27, no. 4 (2012): 461–479.

Haig, David A. "Proximate and Ultimate Causes: How Come? And What For?" *Biology and Philosophy* 28, no. 5 (2013): 781–786.

Haig, David A. "Fighting the Good Cause: Meaning, Purpose, Difference, and Choice." *Biology and Philosophy* 29, no. 5 (2014): 675–697.

Haig, David. *From Darwin to Derrida: Selfish Genes, Social Selves, and the Meanings of Life*. Cambridge, MA: MIT Press, 2020.

Hall, Matthew. *Plants as Persons: A Philosophical Botany*. Albany, NY: SUNY Press, 2011.

Hansson, Sven Ove. "Science and Pseudo-Science." In *Stanford Encyclopedia of Philosophy*, Spring 2017 ed. https://plato.stanford.edu/entries/pseudo-science/.

Huxley, Julian Sorell. *Evolution: The Modern Synthesis*, 5th ed. London: Allen and Unwin, 1948.

Huxley, Thomas Henry. "On the Hypothesis that Animals are Automata, and its History." *Fortnightly Review* 95 (1874): 556–580.

Irigaray, Luce, and Michael Marder. *Through Vegetal Being: Two Philosophical Perspectives.* New York: Columbia University Press, 2016.

James, Jennifer E., Sara M. Willis, Paul G. Nelson, Catherine Weibel, Luke J. Kosinski, and Joanna Masel. "Universal and Taxon-Specific Trends in Protein Sequences as a Function of Age." *eLife* 10 (2021): e57347.

Johnson, Monte Ransome. *Aristotle on Teleology.* New York: Oxford University Press, 2005.

Jones, Derek M. *The Biological Foundations of Action.* New York: Routledge, 2017.

Keller, Evelyn Fox. *Making Sense of Life: Explaining Biological Development with Models, Metaphors, and Machines.* Cambridge, MA: Harvard University Press, 2002.

Koperski, Jeffrey. "Two Bad Ways to Attack Intelligent Design and Two Good Ones." *Zygon* 43, no. 2 (2008): 433–449.

Lamarck, Jean-Baptiste. *Zoological Philosophy: An Exposition with Regard to the Natural History of Animals.* Trans. Hugh Elliot. London: Macmillan, 1914.

Larson, Greger, Philip A. Stephens, Jamshid J. Tehrani, and Robert H. Layton. "Exapting Exaptation." *Trends in Ecology and Evolution* 28, no. 9 (2013): 497–498.

Lenoir, Timothy. *The Strategy of Life: Teleology and Mechanics in Nineteenth Century German Biology.* Chicago: University of Chicago Press, 1982.

Lennox, James G. "Darwin Was a Teleologist." *Biology and Philosophy* 8, no. 4 (1993): 409–421.

Lennox, James G. "The Darwin/Gray Correspondence 1857–1869: An Intelligent Discussion about Chance and Design." *Perspectives on Science* 18, no. 4 (2010): 456–479.

Locke, John. *The Works of John Locke in Nine Volumes,* 12th ed., vol. 1. London: Rivington, 1824.

Lovejoy, Arthur O. *The Great Chain of Being.* Cambridge, MA: Harvard University Press, 1971.

Mayr, Ernst. "Cause and Effect in Biology." *Science* 134, no. 3489 (1961): 1501–1506.

Mayr, Ernst. "The Multiple Meanings of Teleological." In *Toward a New Philosophy of Biology,* edited by Ernst Mayr, 38–66. Cambridge, MA: Harvard University Press, 1988.

Mayr, Ernst. "The Idea of Teleology." *Journal of the History of Ideas* 53, no. 1 (1992): 117–135.

Meyer, Gitte. "In Science Communication, Why Does the Idea of a Public Deficit Always Return?" *Public Understanding of Science* 25, no. 4 (2016): 433–446.

McShea, Daniel W. "Three Trends in the History of Life: An Evolutionary Syndrome." *Evolutionary Biology* 43, no. 4 (2016): 531–542.

Midgley, Mary. *Evolution as Religion*. New York: Routledge, 1985.

Midgley, Mary. *Are You an Illusion?* New York: Routledge, 2014.

Miller, Elaine P. *The Vegetative Soul: From Philosophy of Nature to Subjectivity in the Feminine*. Albany, NY: SUNY Press, 2002.

Millikan, Ruth Garrett. *Language, Thought, and Other Biological Categories*. Cambridge, MA: MIT Press, 1984.

Millstein, Roberta L. "Are Random Drift and Natural Selection Conceptually Distinct?" *Biology and Philosophy* 17, no. 1 (2002): 33–53.

Mix, Lucas J. "Defending Definitions of Life." *Astrobiology* 15, no. 1 (2015): 15–19.

Mix, Lucas J. "Distinguishing Biological Trends from Adaptation." *Philosophy, Theory, and Practice in Biology*. 14 (2022): 10.

Mix, Lucas J. *Life Concepts from Aristotle to Darwin: On Vegetable Souls*. New York: Palgrave, 2018.

Mix, Lucas J. *Life in Space: Astrobiology for Everyone*. Cambridge, MA: Harvard University Press, 2009.

Mix, Lucas J. "Nested Explanation in Aristotle and Mayr." *Synthese* 193, no. 6 (2016): 1817–1832.

Mix, Lucas J., and Joanna Masel. "Chance, Purpose, and Progress in Evolution and Christianity." *Evolution* 68, no. 8 (2014): 2441–2451.

Monod, Jacques. *Chance and Necessity*. New York: Vintage, 1972.

Mutumi, Gregory L., David S. Jacobs, and Henning Winker. "The Relative Contribution of Drift and Deletion to Phenotypic Divergence: A Test Case Using the Horseshoe Bats *Rhinolophus simulator* and *Rhinolophus swinnyi*." *Ecology and Evolution* 7, no. 12 (2017): 4299–4311.

Nagel, Ernest. "Goal-Directed Processes in Biology." *Journal of Philosophy* 74, no. 5 (1977): 261–279.

Nealon, Jeffrey T. *Plant Theory*. Redwood City, CA: Stanford University Press, 2015.

Neander, Karen. "The Teleological Notion of 'Function.'" *Australasian Journal of Philosophy* 69, no. 4 (1991): 454–468.

Nida-Rümelin, Martine. "Doings and Subject Causation." *Erkenntnis* 67, no. 2 (2007): 255–272.

O'Connor, Timothy. *Persons and Causes: The Metaphysics of Free Will*. Oxford: Oxford University Press, 2002.

O'Connor, Timothy. "Conscious Willing and the Emerging Sciences of Brain and Behavior." In *Downward Causation and the Neurobiology of Free Will*, edited by Nancey Murphy, George Ellis, and Timothy O'Connor, 173–186. Berlin: Springer, 2009.

Osler, Margaret J. *Divine Will and the Mechanical Philosophy*. New York: Cambridge University Press, 1994.

Osler, Margaret J. *Reconfiguring the World: Nature, God, and Human Understanding from the Middle Ages to Early Modern Europe*. Baltimore: Johns Hopkins University Press, 2010.

Papineau, David. "Philosophy of Science." In *Blackwell Companion to Philosophy*, 2nd ed., edited by Nicholas Bunnin and Eric P. Tsui-James, 286–316. Hoboken, NJ: Blackwell, 2003.

Papineau, David. "Naturalism." *In Stanford Encyclopedia of Philosophy*, Winter 2016 ed. Stanford University, 1997–. https://plato.stanford.edu/entries/naturalism/.

Pennock, Robert T. "Naturalism, Evidence and Creationism: The Case of Phillip Johnson." *Biology and Philosophy* 11, no. 4 (1996a): 543–549.

Pennock, Robert T. "Reply: Johnson's Reason in the Balance." *Biology and Philosophy* 11, no. 4 (1996b): 565–568.

Perlman, Mark. "The Modern Philosophical Resurrection of Teleology." *Monist* 87, no. 1 (2004): 3–51.

Pittendrigh, Colin S. "Adaptation, Natural Selection and Behavior." In *Behavior and Evolution*, edited by Anne E. Roe and George G. Simpson, 390–416. New Haven, CT: Yale University Press, 1958.

Reiss, John H., and John O. Reiss. (2005) "Natural Selection and the Conditions for Existence: Representational vs. Conditional Teleology in Biological Explanation." *History and Philosophy of the Life Sciences* 27, no. 2 (2005): 249–280.

Rey, Roselyne. "Psyche, Soma, and the Vitalist Philosophy of Medicine." In *Psyche and Soma: Physicians and Metaphysicians on the Mind-Body Problem from Antiquity to Enlightenment*, edited by John P. Wright and Paul Potter, 255–265. Oxford: Clarendon, 2000.

Rosenberg, Alex, and Daniel W. McShea. *Philosophy of Biology: A Contemporary Introduction*. London: Routledge, 2007.

Ruse, Michael. *Monad to Man*. Cambridge, MA: Harvard University Press, 1996.

Ruse, Michael. "Evolutionary Biology and the Question of Teleology." *Studies in the History and Philosophy of Science C* 58 (2016): 100–106.

Schloss, Jeffrey P. "Evolutionary Ethics and Christian Morality: Surveying the Issues." In *Evolution and Ethics: Human Morality in Biological and Religious Perspective*, edited by Phillip Clayton and Jeffrey P. Schloss JP, 1–24. Grand Rapids, MI: Eerdmans, 2004.

Scott, Eugenie C. *Evolution vs. Creationism*, 2nd ed. Berkeley: University of California Press, 2009.

Shipley, Lisa A., Jennifer S. Forbey, and Ben D. Moore. "Revisiting the Dietary Niche: When is a Mammalian Herbivore a Specialist?" *Integrative and Comparative Biology* 49, no. 3 (2009): 274–290.

Simpson, George G. "The Nonprevalence of Humanoids." *Science* 143, no. 3608 (1964): 769–775.

Sloan, Phillip. "Evolutionary Thought Before Darwin." In *Stanford Encyclopedia of Philosophy*, Summer 2017 ed. Stanford University, 1997–. https://plato.stanford.edu/entries/evolution-before-darwin/.

Smith, Felisa A., et al. "Body Size Evolution Across the Geozoic." *Annual Review of Earth and Planetary Sciences* 44 (2016): 523–553.

Sober, Elliott. *Philosophy of Biology*. New York: Oxford University Press, 1993.

Sober, Elliott. *Evidence and Evolution*. Cambridge, UK: Cambridge University Press, 2008.

Ulett, Mark A. "Making the Case for Orthogenesis: The Popularization of Definitely Directed Evolution (1890–1926)." *Studies in the History and Philosophy of Science C* 45 (2014): 124–32.

Walsh, Denis M. "Teleology." In *Oxford Handbook of Philosophy of Biology*, edited by Michael Ruse, 113–137. New York: Oxford University Press, 2008.

Walsh, Denis M., and Andrew Ariew. "A taxonomy of Functions." *Canadian Journal of Philosophy* 26, no. 4 (1996): 493–514.

Wöhlleben, Peter. *The Hidden Life of Trees*. Vancouver, BC: Greystone, 2016.

Wolf, Yuri I., Mikhail I. Katsnelson, and Eugene V. Koonin. "Physical Foundations of Biological Complexity." *Proceedings of the National Academy of Sciences* 115.37 (2018): E8678–E8687.

Wright, Larry. "Functions." *Philosophical Review* 82 (1973):139–168.

CHAPTER 8

Genes and Natural Selection Finalize Nature

Abstract The power, form, and function of living things depend on nutrition and reproduction, which in turn depend on the unique power, form, and function of living things. This etiological recursion, discussed by Aristotle and Kant, was grounded in discrete, dynamic populations by Darwin and the Modern Synthesis. Evolutionary theory naturalizes teleology and finalizes nature with genes and natural selection. Genes store information and grant function, but only as instantiations of a continuous stochastic process described by population genetics. The power aspect of biological agency has shifted to a larger system, a material and energetic feedback loop linking a Darwinian population and the environment it inhabits. Attempts to reify either genes or natural selection independently fail because they miss the recursive quality of biological explanation. Genes are inappropriately teleological when we grant them power, and natural selection is inappropriately teleological when we attribute essence or ends. Only together do they become empirically, conceptually, and mathematically tractable. Although genes and their biochemical milieu are essential to the process—as currently observed—other units of replication and regulation can serve as biological agents in the etiological recursion. This leads to an empirically grounded biological nominalism.

Keywords Adaptation • Gene • Definition of life • Natural selection • Teleology

L. J. Mix, *The End of Final Causes in Biology*,
https://doi.org/10.1007/978-3-031-14017-4_8

Evolutionary theory naturalized teleology, but it also finalized nature. The power, form, and function of living things depend on a process of nutrition and reproduction, which in turn depends on the unique power, form, and function of living things. Aristotle identified this cyclic dynamism 24 centuries ago, but his readers could not grasp the recursive quality of his definitions. They sought static, subsistent entities to underwrite the process, eternal forms with the potential to act even when not acting.

Enlightenment thinkers rejected such abstract entities, looking instead to a fantastic menagerie of biological "agents" as loci for teleology. Most failed. They were excluded from natural science a priori for lacking lawful regularity or a posteriori as agents of a positive directional change. Genes survived as the repositories of form and function, but they did so in a curious way. They resurrected the recursive approach of Aristotle, grounding the explanatory cycle in a physical process of inheritance, variation, and selection.

Genes store form (a.k.a. information) and grant function, but only as instantiations of the continuous stochastic process described by (though not necessarily limited to) population genetics. They lack the power and foresight once attributed to vegetable souls and vital particles, but they retain the formal and final cause aspects, when understood pragmatically as difference makers and "that for the sake of which" accounts. Genes remain relevant as a way of keeping score in a long-term game of fitness.

The power aspect of biological agency shifted to a larger system. It was no longer attributed to an ontologically distinct "agent." The ability to do work in nutrition, development, and reproduction resides in the system itself, a material and energetic feedback loop linking a Darwinian population and the environment it inhabits. Often personified as "natural selection"—though equivalent to the cohort of inheritance, variation, and selection—it took on the agential (efficient cause) character of vegetable souls and "Lamarckism."[1] The system has power yet still lacks the foresight and the other mental properties which led Enlightenment philosophers and Modern biologists to reject "teleology."

Evolutionary theory naturalizes teleology and finalizes nature with genes and natural selection. Attempts to reify either one independently fail because they miss the recursive quality of biological explanation. Genes are inappropriately teleological when we grant them powers or

---

[1] Recall that "Lamarckism" is a nineteenth-century label for environmental drivers, only loosely connected to the position of Lamarck. See Chap. 5.

faculties—particularly the faculty of will, inherent in "selfishness." Natural selection is inappropriately teleological when we attribute essence or ends to the whole system. Only together do they become empirically, conceptually, and mathematically tractable.

Historical theories were as metaphorical as modern ones and modern ones may be less innocent than we hope. Real progress has been made, not through the elimination of teleological language but by grounding that language in an observable process. C. S. Lewis (1964, p. 93) pointed out that the modern description of masses "obeying the law" of gravity is far more anthropomorphic than the Medieval description of earth "inclining" toward the center of the universe. Obeying the law requires both knowledge and will. The new theory is better, not because it is less teleological in expression, but because it is more explicit and useful. In biology, the self-perpetuation of vegetable souls has been replaced with linguistic (or computational) metaphors such as encoding, transcription, and translation. Once again, the new metaphor is more mental and more anthropomorphic and yet better because it is more explicit and precise.

We can speak of genes "directing" metabolism and nature "selecting" traits because we know that the interaction between the two substantiates the metaphor. This recursive foundation for teleology makes it important for biologists to understand exactly how evolutionary theory underwrites the language of power, form, and function. Genes and natural selection can justify teleological language while avoiding the pitfalls of supernatural, progressive, and prospective agency. They work by appealing to an etiological recursion (of form and function) grounded in a physical cycle of form informing matter through metabolism.

ETIOLOGICAL RECURSION

Biological function is no more and no less than the perpetuation of biological forms through time. And biological form is no more and no less than the inherited structure (behavioral as well as physical) of that function. This results an etiological recursion, in which form and function co-evolve. Each explains and validates the other.

Aristotle described the etiological recursion as a confluence of efficient and formal causes.[2] A vegetable soul represents both the author and end of itself through the processes of nutrition and reproduction. Vegetable

---

[2] Chapter 2; Mix (2018, pp. 55–66); see Aristotle's *On the Soul*.

souls produce vegetable souls for the sake of producing vegetable souls. Nutrition is a continuous process (in action and in fulfillment) of form informing matter as the soul gives purpose to atoms, directing them to the common end (organism). The soul does not enable nutrition; it is nutrition. The process defines the form it perpetuates.

Kant also invoked an etiological recursion when he spoke of living things as physical ends, when a thing is "both cause and effect of itself." [3] Final causes in biology can be viewed diachronically as a nexus of efficient causes, a historical sequence of events. We cannot know such ends by pure reason or by observation and yet we must attribute them to living things to understand what they are and what they do. We must treat living things as having ends of their own and of being ends in themselves, not simply means to the ends of another being.

Darwin spelled out the etiological recursion in more detail with the mechanics of populations changing in time. [4] Common ancestry ("the propinquity of descent") and adaptation produce the characters that define each species. Inherited characters, in turn, provide the population from which selection occurs. Darwin revolutionized the concept, however, by recognizing species as pragmatic, nominal categories rather than eternal, ideal forms. His populations changed in time.

Like Lamarck before him, Darwin emphasized the agency of the environment in effecting that change. Unlike Lamarck, Darwin focused on populations shifting—how species originate and change, how characters vary within groups—rather than focusing on individuals acquiring traits. Traits and functions vary within a population. And the trajectory of adaptation in a single lineage can only be understood as part of a larger system, a branching tree. It was neither linear nor normative.

We cannot establish final causes in biology by pure reason or by observation because they are neither ideal nor natural. They are conventions we create to describe a dynamic process of evolution (in action and in fulfillment). The process has no discrete parts, but evolutionary theory can provide conceptual and mathematical indices, by which we understand consistencies and track change through time. These indices are the modern biological agents, genes and organisms.

---

[3] Chapter 4; Mix (2018, pp. 163–164); see Kant's *Critique of Judgment*, part 2, sections 63–68.
[4] Chapter 4; Mix (2018, pp. 199–205); see Darwin's *On the Origin of Species*.

Modern biological agents are not static or subsistent, though we are constantly tempted to think of them as such. They are necessarily pragmatic and inherently relational. They are inextricably intwined with other agents in populations and environments. They are historically and physically situated within a metabolic milieu defined by genes and the biochemical processes that surround them.[5]

PHYSICAL CYCLING

The most successful definitions of life point to two physical processes, metabolism and evolution, one or both of which must be present to call a thing alive (Mix 2015).[6] Metabolism describes a biochemical process of enzyme-mediated reactions, by which an organism maintains itself in the face of increasing entropy. Modern chemistry reveals a far more intricate process than the one Aristotle wrote about (nutrition: the assimilation of elements) but it still describes form informing matter. Metabolism requires continuous energy expenditure at the cellular and subcellular level. Each carbon atom must be acquired, each calorie of energy harvested. Biological actors that mediate this process have been called regulators, of which cells and organisms are the most commonly invoked.

Evolution describes a more periodic process of biological agents being copied imperfectly and, through inheritance, variation, and selection, adapting to their environment. Such agents have been called replicators, of which genes are most often invoked. On Earth, replication cannot occur without regulation. We find genes in the context of cells and organisms. The persistence of genes through generations requires metabolism because of the energy needed to maintain them—preventing and repairing mutations, sequestering and segregating sequences, alongside a variety of epigenetic modifications. Neither can metabolism—as currently understood—occur without genetic information to direct it.

These linked physical cycles provide a necessary foundation for the etiological recursion. Form and function explain each other, but both require

---

[5] Examples of such pragmatic thinking can be found in Godfrey-Smith's (2009) evolutionary nominalism, Haig's (2012, 2020) strategic genes, Walsh's (2015) system of affordances, and Mix's (2018, pp. 232–236) biological nominalism.

[6] Godfrey-Smith (2009) discusses cases where the two result in competing concepts of individuality.

energy to operate. That energy is provided by a metabolic and evolutionary system directed by genes.

Metabolic Power

Metabolism captures energy from the environment and stores it statically as potential energy (electrical energy in ion gradients, chemical energy in reduced chemical species such as ATP and NADPH) and dynamically as heat (Mix 2009, pp. 178–244). This provides the power to maintain dynamic consistency. It enables form and function as well as the imperfect inheritance that allows them to adapt.

One of the most interesting and economically significant properties of life is the production of highly concentrated, highly reduced hydrocarbons that can be used as fuel (e.g., carbohydrates, wood, coal, oil, and gas). Life produces these resources and may, in turn, feed on them and yet, without a Darwinian population, they are not alive.

Aristotle said that a dead body was only a body, a distinct biological entity, equivocally. A biological agent must be dynamic, in action and in fulfillment. This critique reveals more than the inert passivity of a dead body. It reveals the need for ongoing nutrition, not-self transformed into self. Potential energy can be static, but power cannot. It must be observed in change. There must be food being assimilated, resources being consumed, and space being filled.

Similarly, a meaningless word is only a word equivocally. A unit of information must be in the process of communication from one agent to another (even if that process is mistaken or incomplete). "Mindless interpreters" may exist, but not agent-free interpreters. The act of interpretation requires an "agent" for whom and by whom interpretation occurs. Nor can a passive, mechanical molecule contain information without a biological system to transmit or receive it: an alphabet and a language by which it signifies.[7]

This limitation can never be removed. A river is a river, unless frozen, in which case it is just a long, thin block of ice. For an organism to be

---

[7] Mechanical definitions of "information" may be proposed, purely as a measure of entropy if the physical system is well enough known, but Shannon entropy and other semantic definitions require a reference alphabet. Nor can the non-semantic forms of "information" do the necessary work of inheritance. The passage of traits from generation to generation requires shared mechanisms and protocols for encoding and decoding. It also usually involves cellular continuity—shared ancestry, if not linear descent.

organismal, for it to be organized, it must be actively informing something, transforming matter, and molding environment. Metabolic power is necessarily contingent upon a material and energetic environment. Modern science represents a final triumph for Heraclitus' flux. And still, in the great flow of the universe toward disorder, life creates small eddies of form and function.

Genetic Power

Genes have almost no power in this sense and, therefore, differ markedly from previous biological actors. While some potential energy may be attributed to information content—and this is a contentious claim—the power needed to express and interpret that information remains critical to evolutionary teleology. The energy for polymerases, transcriptases, and ribosomes is not stored in polynucleotides, but in mitochondrial energy gradients and reduced biomolecules. Energetically, genes are mechanically passive and not relevant difference makers.[8]

Gene tokens, individual polynucleotide sequences, lack the potential energy to drive reactions. They contain no information in a static or essential way. An alien intelligence discovering a lone polynucleotide could not recognize it as a gene sequence, a packet of information. If it were long enough, they might recognize it as a fuel, a carbon molecule so reduced and improbable that it required life to generate it. They could not recognize the message without a cipher or system of interpretation.

Meanwhile, gene types, strategic genes, and genomes remain abstract entities invoked to describe a dynamic process. As such, they also lack physical power and cannot be agents in this sense. They, too, are manipulated by the larger system.

The power aspect of biological agency did not move to genes. This disruption, I suspect, undergirds the common discomfort with the use of agential language for genes. They can be selfish in function; they can perpetuate their form without perpetuating the form of adjacent genes, of the

---

[8] For this reason, genes cannot be ultimate efficient causes in the language of Aristotle or the Enlightenment. In evolutionary biology, they are treated as proximal efficient causes with natural selection as a more distal efficient cause.

genome, cell, organism, and species in which they occur. They cannot be selfish in will or action. Genes cannot express themselves.[9]

Organismal Power

Walsh (2015) makes a compelling case for organismal agency in regulating complex systems and mediating replicator-environment interactions. "Genes are not units of phenotypic control. Phenotypic control is spread throughout the entire gene/genome/organism/environment system. It is orchestrated and regulated by complex adaptive systems...capable of sensing, adapting, and coordinating responses to their conditions" (p. 132).

Walsh emphasizes his resistance to ontological separation of organism and environment. He cites two prominent biologists from the mid twentieth century (quoted on p. 172): C. H. Waddington claimed that "organism and environment are not two separable things," while Richard Lewontin argued that "There is no organism without an environment, but there is no environment without an organism." Walsh turns, however, to a twentieth-century psychologist when developing his own theory.

J.J. Gibson's theory of ecological psychology shifted forms from internal mental states to ecological *affordances*, shaped by the interaction of animal and environment. Animals do not passively receive information via perception, they actively participate in the meaning-making of a larger, biological system. Thus, Gibson attempted to break the Enlightenment dualism which places formal causes solely in the agential realm. He brought forms into the natural world.

Walsh uses *affordances* in a similar fashion to rescue the agency of organisms. In his opinion, the Modern Synthesis rendered organisms as passive objects, shaped by their environment. Like Gibson, he wants to bring agency over into the natural world. Biological power, form, and function, he argues, reside in the larger teleological—and ecological—system.

Walsh wants to identify organisms as uniquely agential in responding to and shaping the environment (p. 219). Unabashedly agential and teleological in his language, he argues for an emergence of power at the

---

[9] In biological language, some selfish genetic elements "cause" themselves to be expressed (e.g., viruses and plasmid addiction systems), but they lack the power of autopoiesis and must wait for a cell to activate them.

organismal level for serving organismal ends. For example, he claims that organisms repair genes; genes do not repair themselves (p. 204).

It is hard not to interpret this as organisms repairing genes for the sake of organismal ends. But we have seen that this is a difficult claim to make. Cells frequently copy selfish genetic elements at the expense of cellular fitness (dramatically in the case of viruses). Gene tokens may not have power or ends, but strategic genes do. Walsh argues that organisms have goals, affordances, and a flexible repertoire of behaviors. I would argue that strategic genes have all three, but I can see where Walsh and others might see that as a stretch. Walsh does attribute similar agency to genomes and gene regulatory networks.

Overall, I am uncomfortable with the psychological and organism-focused language in both Gibson and Walsh. If the emergence of agency extends across a variety of biological actors at a variety of scales (including viruses, plasmids, organelles, and endosymbionts), I believe it is consistent with the picture I present here. If it creates a new ontological entity, of which these other biological agents are merely components, it strikes me as unnecessarily metaphysical.

Notably, Walsh claims that his approach to bioteleology is wholly natural and empirical while not being mechanical in a Cartesian sense (p. 222). He continues in a long tradition of renegotiating the bounds of nature. I encourage interested readers to follow his argument closely and come to their own conclusions.

I wholeheartedly support the emphasis on agent/environment interaction and reciprocal formation of power, form, and function where the biological agent is defined broadly and pragmatically. In that context, I see form and function as irreducibly contextual and yet clearly connected to a specific agent. This sequence defines a given strategic gene. That genotype defines a given individual organism. This gene is the "for whom" of a given phenotype. That organism is the "for whom" of a certain behavior. I am less confident in the attribution of power to particular agents. I think this reflects an ill-advised analogy from the conscious human to all biological agents. To the extent that power arises from the reciprocal relationship, it is a relationship between all things, living and non-living, and not localizable to any singular agent—at least within the paradigm of methodological naturalism.

Systemic Power

Two examples reveal the importance of considering power at the population/environment level, colony structure and gene transfer agents (GTAs). In colonies, the group can be said to "act" on individuals by regulating nutrition and reproduction. Not only do individuals act to direct or restrain one another (e.g., the selective feeding of bee larvae), but the density of individuals and the concentration of chemical signals can also impact nutrition and reproduction (e.g., quorum sensing in slime molds). Colony action cannot be reduced to the sum of individual actions. It is density dependent and, therefore, time and space dependent.

Some readers will consider colonies analogous to individual organisms, at least in part, and conclude that the organism is still the primary unit of agency. The reality, however, remains more bizarre. GTAs have been discovered in diverse prokaryotes. Apparently domesticated bacteriophages, GTA genes in the host genome incorporate some, but not all, of the GTA genes along with other sequences into protein coats before lysing the cell. These genetic lifeboats attach to other cells and offload the genes. GTAs may be involved in the extensive breadth of metabolic capability among the alpha and beta proteobacteria, including the wide dispersal of phototrophy. They "function" at the class and phylum level.

Genetic and metabolic networks defy easy characterization. Biological power appears across scales, from the catalytic agency of enzymes to the information capacity of plasmids and GTAs, the chemical potential of plastids, the regulatory power of cells, the homeostasis of plants and animals, and the social power of colonies. Global chemical cycles may reveal even broader regulation, as in the complex relationship between vegetation, carbon dioxide, planetary albedo, and climate. All of these represent genuine biological form and function.

A sufficiently broad definition of "organism" might be constructed that would equate it with any regulatory unit up to and including the biosphere (e.g., Doolittle 2014). Discussions of Gaia move in precisely this direction. And yet, it is harder to imagine how the etiological recursion would occur at this level, as the characterization of inheritance, variation, and selection become more difficult without a super-organismal population in which selection occurs.

The common-sense definition of organism as a discrete metabolic and reproductive entity cannot be the exclusive or even primary locus of biological power. It always and necessarily exists in dynamic equilibrium with a population and an environment. Power resides in a continuum of complexity ranging from genes to the biosphere, with the form and function of biological agency just a little less clear at the ends.

> Biology as natural science should treat individuals (organisms, species, units of selection, and units of inheritance) as arbitrary names, useful to the extent that they allow us to make accurate predictions within the framework of evolutionary theory. Key to biological nominalism will be the idea that claims of organismality are not exclusive. One organism may overlap another. This represents a departure from traditional life-concepts precisely in our use of formal causes. (Mix 2018, p. 225)

Environmental Action

Form informing matter is not (following Plato) a one-way process, but (following Aristotle and Darwin) a continuous recursive one. This is the problem of trying to speak about evolutionary forces as though they could be abstracted from material contingency. The forces do not exist without the populations they act upon.

This leaves us with a challenge for understanding the power aspect of teleology. The divine will of Medieval theology and the Spirit of German Idealism failed as biological theories. Natural selection must be significantly different, and here the three constraints on bioteleology do real work.

Methodological Naturalism

Natural selection must be lawfully regular. Biologists can and should change the ways we characterize that regularity, but we must always characterize it as regular. Natural selection will always act in the same way. Current understanding rests on the foundations of the Modern Synthesis and population genetics. Natural selection occurs as a stochastic fluctuation of allele frequencies within a real population. While I am, in theory, amenable to an expanded understanding of life, I will remain skeptical until such a time as it can be mathematically described and concretely demonstrated.

The abstractions of cultural and mimetic evolution, though compelling, are not mathematically tractable because they lack discrete units of inheritance and selection. It may be that such units will be identified in the future, as polynucleotide genes were identified a century ago. They have not been yet. Similarly, the well-characterized process of chemical evolution cannot be operationalized as natural selection because there is no pre-existing system to provide the power and dynamic consistency necessary for adaptation.

It has become popular to speak of a natural progression from chemical evolution through genetic evolution to cultural evolution. While this makes intuitive sense, it lacks conceptual power because it loses the explicit connection between the etiological recursion and a physical cycle.

Prior to the first Darwinian population, there was no energy repository to "drive" increases in complexity, the generation of new form and function. Chemical evolution occurs in the same way as stellar and planetary evolution. All three describe change through time. In the context of natural science, they are lawfully regular and may even result in small increases in complexity (e.g., lipid vesicle formation and layer formation in stars). They lack the lawful regularity specific to natural selection, namely, form informing matter through inheritance, variation, and selection formalized as a stochastic fluctuation of alleles in a real population. They lack the open-ended teleology of organisms.

Similarly, in the mental space of cultural evolution, there is no potential energy to do work. We often miss this because the metabolic power of genetic evolution literally and continuously fuels the conceptual space of cultural evolution and the code space of artificial life. Insofar as either exists, they do so as an extension of genetic evolution, not as a new regime.[10]

With all previous generations of biologists, we want to explain the continuity of physical, physiological, and psychological phenomena. The founders of natural science set up the bounds of lawful regularity in a way that made physics and physiology tractable, but we must continue to observe those bounds if we want natural science to succeed.[11] At present, natural selection does not quite stretch to physics, because we do not

[10]We might imagine an ideal artificial intelligence which could operate solely on solar power and thus transcend the physical cycle which produced fossil fuels. It would, nonetheless, derive its form and function from the etiological recursion of human life. An intelligence not based on this recursion would not be artificial. And a non-human "natural" intelligence of this kind has not been observed.

[11]One of those bounds was to place psychology out of bounds. But let us assume that the Renaissance psychology of interiority, (Platonic) intellect, and spirituality might be replaced with a phenomenological psychology.

know how the physical cycle started. Nor does it stretch to psychology because mental phenomena (e.g., consciousness and reason) cannot be described with lawful regularity.

Excellent research programs exist. I have hope that science can bridge these gaps. Evolutionary theory has not done so yet. The etiological recursion and physical cycle justify a coherent, if pragmatic, theory of power, form, and function in biology and should be retained until a better—and equally lawful—alternative presents itself.

It seems tedious to place so many qualifiers on natural selection, but those qualifiers are necessary because they ground it empirically. Attempts to label the origin of life and the origin of consciousness as natural selection pull biology back into metaphysics in unproductive ways. "Evolution" alone will not do, just as vegetable souls would not do. We need the biochemical details of how change occurs, how form and function arise, and where the power to do this work comes from. [12]

Local Adaptation

Natural selection is inescapably local because it is, by definition, grounded in a specific time and place through the interaction of life and environment. Attempts to abstract biology and definitions of life from physical details will always be thwarted by this contingency. Astrobiologists describe this as the $N = 1$ problem. Currently we have a sample size of one (instance of life, that is Earth life), which limits our theory and constrains the search for life elsewhere.

It is easy to get caught up in the challenges this raises, but I wish to emphasize the opposite point. Life on Earth represents a concrete and amazingly consistent data set. The ubiquity of nucleic acid gene, amino acid protein, lipid-bounded vesicle biochemistry reveals something important about life on Earth. It may be that the origin of life is vanishingly rare. Or, once established, our particular instantiation of life may have been efficient enough to outcompete other instantiations. In either case, known life is robust and surprisingly uniform.

---

[12] Other physical cycles can produce chaotic systems and even emergent complexity. My argument is not that lawful regularity prevents innovation. Rather, the lawful regularity *specific to natural selection* is an etiological recursion dependent upon a physical cycle. Natural selection may well result from another lawfully regular process. It may produce another lawfully regular process. But, unlike vegetable souls and autocatalysts, it cannot produce itself.

Natural selection rewards those traits that increase the power of the physical cycle. Despite many attempts to synthesize life in the lab, only the nucleic acid gene, amino acid protein, lipid-bounded vesicle biochemistry evolves in this open-ended way. It is worth imagining, looking for, and trying to make other such physical cycles, but until one appears natural selection will be Earthed in one singular biochemistry.

The biochemistry of Earth has a characteristic background energy. In one sense, drift is noise; it obscures the signal of adaptation. In another sense, it represents a real process of dynamic flux necessary to maintain the system (Mix 2022). As Sober (2008, pp. 192–199) and Millstein (2000, 2002) point out, the options are not selection or drift, but selection and drift or drift alone. Stochasticity is an essential feature of the physical system and a manifestation of the power trapped therein.

This should be celebrated as practical knowledge. Rationalist approaches to defining life on first principles have been provocative, but empiricist approaches to characterizing natural selection have been more productive. Instead of asking "what is life?" abstractly, we can ask "what is it about natural selection on Earth that makes it interesting?" Clearly, the power aspect catches our attention. As do the adaptations, forms, and functions it provides.

Blindness

Biological agency cannot be prospective, because there is no entity to view or think about the future. Genes, (most) organisms, and populations lack the mental faculties necessary to hold any representations, whether perceived (sensation) or imagined (reason). They are not even objective entities, only pragmatic indices for describing a continuous flux of matter and energy. Nor can the flux itself have intentions or foreknowledge. It is neither more nor less than energy caught in a system.

When a cellular metabolism acts on the polynucleotide we call it "interpretation" because we have, pragmatically, reified parent and offspring as biological agents talking to each other. We attribute mental states to both and say one communicates an abstract form to the other by way of a signal or message. This metaphor draws heavily on psychological concepts. And it can be quite useful, so long as we remember that neither agents nor message exist on their own. A ribosome cannot "interpret" a gene without a human to interpret the events in this way.

We can never know what the correct message is supposed to be. Neither sender nor receiver intended accurate communication because neither subsists and neither has prospect. This proves especially important when considering the question of adaptation. The function encoded by a gene arose because it was adaptive and heritable. It persists because it is adaptive and heritable. Yet it may be adaptive in multiple ways at different times, in different environments, or for different agents. We should always ask whether another story, with different agents, messages, and adaptations, might fit the data just as well as the one we started with.

Adaptations, like agents, are pragmatic stories, dependent upon story tellers and narrative context for their meaning. They describe a real change in the world, but their forms and ends are attributed. We observe changes in frequency (though only after binning the data in theory-laden ways). We ascribe units of inheritance and selection.

In most ways, the biologist remains as blind as genes and natural selection. The flux of life explores a phenotype space beyond comprehension (and a genotype space beyond computation). Even if we knew the full range of possible environments, we could not predict the range of actual adaptations. This should make us humble about the direction of adaptation. As with nature itself—for it is nature itself—it is regular enough for us to discern laws, but unpredictable enough that observation will always be necessary.

We do not know how prospective agency (standard human agency) arises, but we can say it arises within the flux. This should be celebrated as practical knowledge as well. Indeed, it was celebrated by Aristotle, Augustine, Aquinas, and many others as an important insight into humanity. Our psychology occurs in the context of our physiology. We might debate whether this is necessary or eternal, but it is clearly normal.[13] Biologists should avoid the ultimate cause questions of divine and human will (as unmoved movers), because they move us out of the flux and into metaphysics. Nonetheless, we can ask some very interesting questions about the population structures and physical environments in which humanity arose.

---

[13] Many readers will be surprised to learn that all three thought humans are necessarily physical and animal. Many, if not most, classical thinkers recognized the embodiment of mind and soul. Modern dualism dates to the Enlightenment, though it has many earlier antecedents (Mix 2018).

Completeness

A final note is warranted on how I have been treating the term "environment." Just as genes and other biological "agents" must be treated pragmatically—being neither ideal nor empirical—so must the environment. It can be easy to drift into a dualism of agent and environment—cell and surroundings (i.e., dividing at the cell membrane) or organism and surroundings (e.g., dividing at the skin, bark, or cell wall) or even species and surroundings (i.e., distinguishing natural from artificial or sexual selection). While each of these pairings has its place, the pragmatic and nominal flexibility of agents calls for a similar flexibility of environments.

The intra-cellular environment may be significant for gene selection, just as the intercellular environment may be important for organismal selection. The twentieth century revealed a plethora of symbiotes intertwined with replicators and regulators at every level. Arguments about selection should be grounded in a concrete awareness of the relevant environment. What is there, other than the replicator or regulator itself, that contributes to its fitness?

Rather than a spatial category, the environment describes the complete set of circumstances relevant to the survival and reproduction of individuals.[14] This might include radioactive elements, chemical poisons, or microorganisms that remain within a cell for multiple generations. Occasionally labeled "epigenetic inheritance," this consistency of material causes is better understood as a property of the environment along with other resources and hazards present in a habitat. In Aristotelian terms, they are relevant to material cause accounts, but not to the efficient/formal/final cause that defines a living being. In more modern terms, they are analogous to atoms and gene tokens—below the level of biological organization. Only once they become integral to the function of an organism (e.g., manganese in oxygen evolution for phototrophs, mitochondria for energy production in eukaryotes) should they be treated as inherited traits of an organism.

The potential for mismatch between organism as replicator and organism as regulator can, likewise, lead to counterintuitive uses of "environment." Natural selection on viroids, plasmids, and other selfish genetic elements (replicators) can occur with a constant regulatory background (e.g., ribosomes and transfer RNAs). Conversely, we can think of natural

---

[14]Walsh (2022, p. 69): "The environment stands proxy for an enormously complex set of conditions—both within organisms and beyond—that tokens of a trait type might encounter."

selection on regulators (e.g., human lineages) in which paired replicators form the background (e.g., mitochondrial and nuclear lineages). Are mitochondria part of the human organism or part of the environment? That depends on the evolutionary question being asked. A nominal biological agent requires us to be constantly aware of what exactly is replicating or regulating in any given model. Only then can we identify the relevant population and environment.[15]

ETIOLOGICAL RECURSION VERSUS PHYSICAL CYCLE

Some readers may be confused by the last two sections. If the etiological recursion requires the physical cycle and the physical cycle requires population genetics, it seems that genes must be the only biological agents of interest. And yet, the last section seems to suggest multi-level selectionism. Which one is it?

This brings us back to the important distinction between form and function on the one hand, provided by the etiological recursion, and power on the other, provided by the physical cycle. The etiological recursion is entirely theoretical, whether in the Aristotelian, Kantian, or Darwinian formulation. Any system with the right features (form informing matter cyclically) would do. It just so happens that we have only observed the power necessary to do this in one type of matter—consistent with population genetics.

Power has been separated from form and function. Thus, biological individuals might be found at multiple levels—relevant difference makers, be they replicators or regulators. Genes will, however, always be a material cause—a compositional element—of those individuals. This gives genes an interesting priority. Natural selection is always occurring at that level, regardless of whether and how it might occur at other levels.

With many in the astrobiology, origin of life, and synthetic biology communities, I look forward to a day when another physical cycle might be found. It may be that natural selection only occurs, and can only occur, in the context of genes. Knowing this would tell us something profound about the universe. Alternatively, finding another physical cycle would

---

[15] Mitochondria are part of human cells and they are organisms when treating cells or larger organisms as the biological agent, the locus of form and function and the target of selection. Mitochondria are part of the environment when treating plasmids, nuclear genomes, or other subcellular units as the agent.

reveal a more fundamental truth about teleology and the possibility for purpose in the cosmos. For now, we can only say that genes provide the basement for such meaning and the biosphere a roof (within the natural sciences) and that natural teleology—as power, form, and function—arises somewhere in between.[16]

* * *

The deep history of biology reveals persistent questions about life. Over two millennia ago, Aristotle identified the key features of organization and dynamic consistency. His theory of life, the vegetable soul, proved more descriptive than explanatory, but it has shaped research agendas to the present day. His distinction of vegetable, animal, and rational activities set the stage for an ongoing challenge. How do we accurately describe humans as operationally unique while recognizing their embodiment in both Earth and biosphere? His naming of material, efficient, formal, and final causes provided categories for epistemological reflection and innovation in the Enlightenment, including the rejection of mentalized formal and final causes.

Enlightenment and Early Modern biologists explored new ways of understanding biological form and function, considering a vast array of biological agents, most unimaginably foreign to modern sensibilities. Kant and Darwin created new ways of thinking about form and function in the context of an etiological recursion. They suggested pragmatic attributions of identity and meaning that captured a real but dynamic flux of form informing matter through time. In the early twentieth century, biology took on a new, coherent epistemic foundation basing form and function concretely in genes and natural selection. Three constraints on teleology enabled this new view of teleology: methodological naturalism, local adaptation, and blind chance.

The power component of evolution underwrites both form and function in living things through a very specific biochemistry. It differentiates evolutionary teleology from earlier views, separating the agency of genes

---

[16] This is a book for biologists. Other epistemologies provide form, function, and meaning in other ways. Limits on biological teleology are not meant to disprove other teleologies. To the contrary, they help biologists and non-biologists alike to identify non-biological teleologies, such as providence and progress, and ask why they may or may not be useful ways of understanding the world.

and organisms from the agency of natural selection and grounding both in an observable process of form informing matter. Without that broader picture, biological form and function lose their meaning. Thus, the biological agent, viewed as metabolic regulator or evolutionary replicator, constantly coevolves with its surroundings and constantly adapts.

Future progress calls for a better understanding of how the etiological recursion occurs within these constraints and a more detailed picture of the stochastic processes underlying natural selection. Which modern biological agents are most useful as units of inheritance, variation, and selection? How do different mathematical frames (e.g., inclusive fitness or kin selection) inform one another? And can we map one onto another reliably (e.g., moving from genetic drift to phenotypic "drift" or from trait selection to gene selection)? A good understanding of bioteleology will help researchers as they design experiments and interpret data in these areas.

More broadly, public communication of science calls for a profound and internalized understanding of how genes and natural selection can naturalize teleology and finalize nature. Teleology matters for biological questions of emergence and convergence as well as social questions about human embodiment, identity, and dignity. Too often, biologists have yielded the public square, allowing others to define what is "natural" and even "evolutionary." These are not the province of biologists alone, but evolutionary biology has something important to offer when informed by the rich history and philosophy of the field as well as the breadth of empirical data. The teleology of biology matters in the end.

REFERENCES

Doolittle, W. Ford. "Natural Selection Through Survival Alone and the Possibility of Gaia." *Biology and Philosophy* 29, no. 3 (2014): 415–423.

Godfrey-Smith, Peter. *Darwinian Populations and Natural Selection*. Oxford: Oxford University Press, 2009.

Haig, David A. "The Strategic Gene." *Biology and Philosophy* 27, no. 4 (2012): 461–479.

Haig, David. *From Darwin to Derrida: Selfish Genes, Social Selves, and the Meanings of Life*. Cambridge, MA: MIT Press, 2020.

Lewis, Clive Staples. *The Discarded Image: An Introduction to Medieval and Renaissance Literature*. Cambridge, UK: Cambridge University Press, 1964.

Millstein, Roberta L. "Chance and Macroevolution." *Philosophy of Science* 67, no. 4 (2000): 603–624.

Millstein, Roberta L. "Are Random Drift and Natural Selection Conceptually Distinct?" *Biology and Philosophy* 17, no. 1 (2002): 33–53.

Mix, Lucas J. *Life in Space: Astrobiology for Everyone.* Cambridge, MA: Harvard University Press, 2009.

Mix, Lucas J. "Defending Definitions of Life." *Astrobiology* 15, no. 1 (2015): 15–19.

Mix, Lucas J. *Life Concepts from Aristotle to Darwin: On Vegetable Souls.* New York: Palgrave, 2018.

Mix, Lucas J. "Distinguishing Biological Trends from Adaptation." *Philosophy, Theory, and Practice in Biology* 14 (2022): 10.

Sober, Elliott. *Evidence and Evolution.* New York: Cambridge University Press, 2008.

Walsh, Denis M. *Organisms, Agency, and Evolution.* Cambridge: Cambridge University Press, 2015.

Walsh, Denis M. "Environment as Abstraction." *Biological Theory* 17 (2022): 68–79.

AUTHOR INDEX[1]

---

[1] Note: Page numbers followed by 'n' refer to notes.

Subject Index[1]

---

[1] Note: Page numbers followed by 'n' refer to notes.